INVESTIGATIONS IN NUMBER, DATA, AND SPACE®

Tables and Graphs

Patterns of Change

Grade 5

Also appropriate for Grade 6

Cornelia Tierney
Ricardo Nemirovsky
Tracy Noble

Developed at TERC, Cambridge, Massachusetts

Dale Seymour Publications®

The *Investigations* curriculum was developed at TERC (formerly Technical Education Research Centers) in collaboration with Kent State University and the State University of New York at Buffalo. The work was supported in part by National Science Foundation Grant No. MDR-9050210. TERC is a nonprofit company working to improve mathematics and science education. TERC is located at 2067 Massachusetts Avenue, Cambridge, MA 02140.

This project was supported, in part, by the
National Science Foundation
Opinions expressed are those of the authors and not necessarily those of the Foundation

This book is published by Dale Seymour Publications®, an imprint of the Alternative Publishing Group of Addison-Wesley Publishing Company.

Project Editor: Catherine Anderson
Series Editor: Beverly Cory
Manuscript Editor: Nancy Tune
ESL Consultant: Nancy Sokol Green
Production/Manufacturing Director: Janet Yearian
Production/Manufacturing Manager: Karen Edmonds
Production/Manufacturing Coordinator: Barbara Atmore, Shannon Miller
Design Manager: Jeff Kelly
Design: Don Taka
Composition: Publishing Support Services
Illustrations: Susan Jaekel and Carl Yoshihara
Cover: Bay Graphics

Printed on Recycled Paper

Order number DS21433
ISBN 0-86651-997-1
1 2 3 4 5 6 7 8 9 10-ML-00 98 97 96

Principal Investigator **Susan Jo Russell**
Co-Principal Investigator **Cornelia C. Tierney**
Director of Research and Evaluation **Jan Mokros**

Curriculum Development
Joan Akers
Michael T. Battista
Mary Berle-Carman
Douglas H. Clements
Karen Economopoulos
Claryce Evans
Marlene Kliman
Cliff Konold
Jan Mokros
Megan Murray
Ricardo Nemirovsky
Tracy Noble
Andee Rubin
Susan Jo Russell
Margie Singer
Cornelia C. Tierney

Evaluation and Assessment
Mary Berle-Carman
Jan Mokros
Andee Rubin
Tracey Wright

Teacher Support
Kabba Colley
Karen Economopoulos
Anne Goodrow
Nancy Ishihara
Liana Laughlin
Jerrie Moffett
Megan Murray
Margie Singer
Dewi Win
Virginia Woolley
Tracey Wright
Lisa Yaffee

Administration and Production
Irene Baker
Amy Catlin
Amy Taber

Cooperating Classrooms for This Unit
Jo-Ann Pepicelli
Boston Public Schools, Boston, MA
Barbara Fox
Cambridge Public Schools, Cambridge, MA
These teachers worked in concert with Marlene Kliman, Tracey Wright, David Carraher, and Steve Monk to carry out the research and planning for this unit.

Technology Development
Douglas H. Clements
Julie Sarama

Video Production
David A. Smith
Judy Storeygard

Consultants and Advisors
Deborah Lowenberg Ball
Marilyn Burns
Mary Johnson
James J. Kaput
Mary M. Lindquist
Leslie P. Steffe
Grayson Wheatley

Graduate Assistants
Kent State University:
Richard Aistrope, Kathryn Battista,
Caroline Borrow, William Hunt

State University of New York at Buffalo:
Jeffery Barrett, Julie Sarama,
Sudha Swaminathan, Elaine Vukelic

Harvard Graduate School of Education:
Dan Gillette, Irene Hall

CONTENTS

Teacher Notes

Investigations in Number, Data, and Space® is a K–5 mathematics curriculum with four major goals:

- to offer students meaningful mathematical problems
- to emphasize depth in mathematical thinking rather than superficial exposure to a series of fragmented topics
- to communicate mathematics content and pedagogy to teachers
- to substantially expand the pool of mathematically literate students

The *Investigations* curriculum embodies an approach radically different from the traditional textbook-based curriculum. At each grade level, it consists of a set of separate units, each offering 2–6 weeks of work. These units of study are presented through investigations that involve students in the exploration of major mathematical ideas.

Approaching the mathematics content through investigations helps students develop flexibility and confidence in approaching problems, fluency in using mathematical skills and tools to solve problems, and proficiency in evaluating their solutions. Students also build a repertoire of ways to communicate about their mathematical thinking, while their enjoyment and appreciation of mathematics grows.

The investigations are carefully designed to invite all students into mathematics—girls and boys, diverse cultural, ethnic, and language groups, and students with different strengths and interests. Problem contexts often call on students to share experiences from their family, culture, or community. The curriculum eliminates barriers—such as work in isolation from peers, or emphasis on speed and memorization—that exclude some students from participating successfully in mathematics. The following aspects of the curriculum ensure that all students are included in significant mathematics learning:

- Students spend time exploring problems in depth.
- They find more than one solution to many of the problems they work on.
- They invent their own strategies and approaches, rather than relying on memorized procedures.
- They choose from a variety of concrete materials and appropriate technology, including calculators, as a natural part of their everyday mathematical work.
- They express their mathematical thinking through drawing, writing, and talking.
- They work in a variety of groupings—as a whole class, individually, in pairs, and in small groups.
- They move around the classroom as they explore the mathematics in their environment and talk with their peers.

While reading and other language activities are typically given a great deal of time and emphasis in elementary classrooms, mathematics often does not get the time it needs. If students are to experience mathematics in depth, they must have enough time to become engaged in real mathematical problems. We believe that a minimum of five hours of mathematics classroom time a week—about an hour a day—is critical at the elementary level. The plan and pacing of the *Investigations* curriculum is based on that belief.

For further information about the pedagogy and principles that underlie these investigations, see the Teacher Notes throughout the units and the following books:

- *Implementing the* Investigations in Number, Data, and Space® *Curriculum*
- *Beyond Arithmetic: Changing Mathematics in the Elementary Classroom*

The *Investigations* curriculum is presented through a series of teacher books, one for each unit of study. These books not only provide a complete mathematics curriculum for your students, they offer materials to support your own professional development. You, the teacher, are the person who will make this curriculum come alive in the classroom; the book for each unit is your main support system.

While reproducible resources for students are provided, the curriculum does not include student books. Students work actively with objects and experiences in their own environment and with a variety of manipulative materials and technology, rather than with workbooks and worksheets filled with problems. We also make extensive use of the overhead projector as a way to present problems, to focus group discussion, and to help students share ideas and strategies. If an overhead projector is available, we urge you to try it as suggested in the investigations.

Ultimately, every teacher will use these investigations in ways that make sense for his or her particular style, the particular group of students, and the constraints and supports of a particular school environment. We have tried to provide with each unit the best information and guidance for a wide variety of situations, drawn from our collaborations with many teachers and students over many years. Our goal in this book is to help you, as a professional educator, implement this mathematics curriculum in a way that will give all your students access to mathematical power.

Investigation Format

The opening two pages of each investigation help you get ready for the student work that follows. Here you will read:

What Happens—a synopsis of each session or block of sessions.

Mathematical Emphasis—the most important ideas and processes students will encounter in this investigation.

What to Plan Ahead of Time—materials to gather, student sheets to duplicate, transparencies to make, and anything else you need to do before starting.

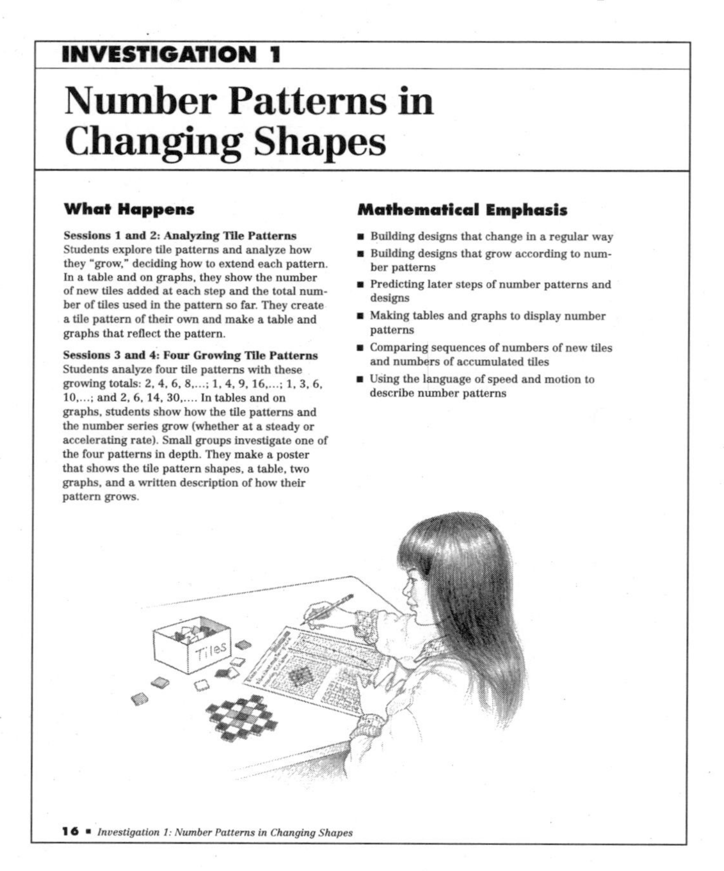

INVESTIGATION 1

Number Patterns in Changing Shapes

What Happens

Sessions 1 and 2: Analyzing Tile Patterns
Students explore tile patterns and analyze how they "grow," deciding how to extend each pattern. In a table and on graphs, they show the number of new tiles added at each step and the total number of tiles used in the pattern so far. They create a tile pattern of their own and make a table and graphs that reflect the pattern.

Sessions 3 and 4: Four Growing Tile Patterns
Students analyze four tile patterns with these growing totals: 2, 4, 6, 8,...; 1, 4, 9, 16,...; 1, 3, 6, 10,...; and 2, 6, 14, 30,.... In tables and on graphs, students show how the tile patterns and the number series grow (whether at a steady or accelerating rate). Small groups investigate one of the four patterns in depth. They make a poster that shows the tile pattern shapes, a table, two graphs, and a written description of how their pattern grows.

Mathematical Emphasis

- Building designs that change in a regular way
- Building designs that grow according to number patterns
- Predicting later steps of number patterns and designs
- Making tables and graphs to display number patterns
- Comparing sequences of numbers of new tiles and numbers of accumulated tiles
- Using the language of speed and motion to describe number patterns

16 ■ *Investigation 1: Number Patterns in Changing Shapes*

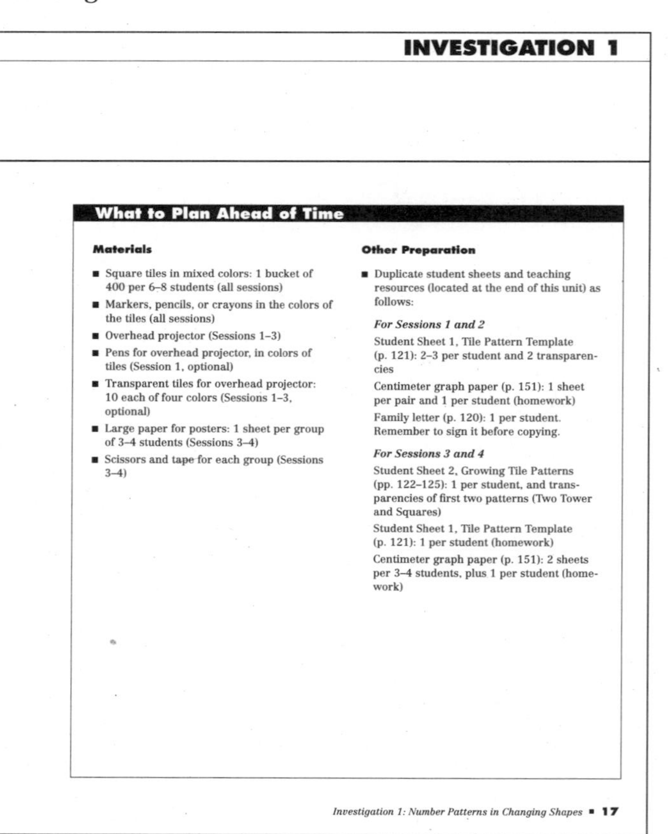

INVESTIGATION 1

What to Plan Ahead of Time

Materials

- Square tiles in mixed colors: 1 bucket of 400 per 6–8 students (all sessions)
- Markers, pencils, or crayons in the colors of the tiles (all sessions)
- Overhead projector (Sessions 1–3)
- Pens for overhead projector, in colors of tiles (Session 1, optional)
- Transparent tiles for overhead projector: 10 each of four colors (Sessions 1–3, optional)
- Large paper for posters: 1 sheet per group of 3–4 students (Sessions 3–4)
- Scissors and tape for each group (Sessions 3–4)

Other Preparation

- Duplicate student sheets and teaching resources (located at the end of this unit) as follows:

For Sessions 1 and 2

Student Sheet 1, Tile Pattern Template (p. 121): 2–3 per student and 2 transparencies

Centimeter graph paper (p. 151): 1 sheet per pair and 1 per student (homework)

Family letter (p. 120): 1 per student. Remember to sign it before copying.

For Sessions 3 and 4

Student Sheet 2, Growing Tile Patterns (pp. 122–125): 1 per student, and transparencies of first two patterns (Two Tower and Squares)

Student Sheet 1, Tile Pattern Template (p. 121): 1 per student (homework)

Centimeter graph paper (p. 151): 2 sheets per 3–4 students, plus 1 per student (homework)

Investigation 1: Number Patterns in Changing Shapes ■ 17

Sessions Within an investigation, the activities are organized by class session, a session being a one-hour math class. Sessions are numbered consecutively through an investigation. Often several sessions are grouped together, presenting a block of activities with a single major focus.

When you find a block of sessions presented together—for example, Sessions 1, 2, and 3—read through the entire block first to understand the overall flow and sequence of the activities. Make some preliminary decisions about how you will divide the activities into three sessions for your class, based on what you know about your students. You may need to modify your initial plans as you progress through the activities, and you may want to make notes in the margins of the pages as reminders for the next time you use the unit.

Be sure to read the Session Follow-Up section at the end of the session block to see what homework assignments and extensions are suggested as you make your initial plans.

While you may be used to a curriculum that tells you exactly what each class session should cover, we have found that the teacher is in a better position to make these decisions. Each unit is flexible and may be handled somewhat differently by every teacher. While we provide guidance for how many sessions a particular group of activities is likely to need, we want you to be active in determining an appropriate pace and the best transition points for your class.

Ten-Minute Math At the beginning of some sessions, you will find Ten-Minute Math activities. These are designed to be used in tandem with the investigations, but not during the math hour. Rather, we hope you will do them whenever you have a spare 10 minutes—maybe before lunch or recess, or at the end of the day.

Ten-Minute Math offers practice in key concepts, but not always those being covered in the unit. For example, in a unit on using data, Ten-Minute Math might revisit geometric activities done earlier in the year. Complete directions for the suggested activities are included at the end of each unit. A compilation of Ten-Minute Math activities is also available as a separate book.

Sessions 1 and 2

Analyzing Tile Patterns

Materials

- Overhead projector
- Transparent tiles for overhead (optional)
- Transparencies of Student Sheet 1
- Overhead pens in colors of tiles (optional)
- Centimeter graph paper (1 per pair and 1 per student, homework)
- Square tiles in mixed colors (1 bucket of 400 per 6–8 students)
- Student Sheet 1 (2–3 per student)
- Markers, pencils, or crayons in the colors of the tiles (optional)
- Family letter (1 per student)

What Happens

Students explore tile patterns and analyze how they "grow," deciding how to extend each pattern. In a table and on graphs, they show the number of new tiles added at each step and the total number of tiles used in the pattern so far. They create a tile pattern of their own and make a table and graphs that reflect the pattern. Their work focuses on:

- predicting later steps in tile patterns and in number patterns
- building designs that change in a regular way
- making tables and graphs that describe growing tile patterns
- finding patterns in step numbers and totals

Activity

Exploring Tile Patterns

To introduce the idea of making a pattern that grows in a regular way, use transparent tiles on an overhead projector or make tile designs on a flat surface where students can gather around. Build the beginning of a Threes Tower—a growing stack of rows of three tiles. Start with one color at the bottom and build up, changing the color for each new row. (Throughout this investigation, we will illustrate patterns with tiles in four colors, red (R), blue (B), green (G), and yellow (Y), in no particular order.)

Y Y Y

As you make the first three rows, talk about what you are doing.

I'm building a pattern. For each new step of the pattern, I'll change the tile color so the step is clear.

R R R
Y Y Y

18 ▪ *Investigation 1: Number Patterns in Changing Shapes*

Activities The activities include pair and small-group work, individual tasks, and whole-class discussions. In any case, students are seated together, talking and sharing ideas during all work times. Students most often work cooperatively, although each student may record work individually.

Choice Time In some units, some sessions are structured with activity choices. In these cases, students may work simultaneously on different activities focused on the same mathematical ideas. Students choose which activities they want to do, and they cycle through them.

You will need to decide how to set up and introduce these activities and how to let students make their choices. Some teachers present them as station activities, in different parts of the room. Some list the choices on the board as reminders or have students keep their own lists.

Excursions Some of the investigations in this unit include *excursions*—activities that could be omitted without harming the integrity of the unit. This is one way of dealing with the overabundance of fascinating mathematics to be studied—much more than a class has time to explore in any one year. Excursions give you the flexibility to make different choices from year to year. For example, you might do the excursion in *Patterns of Change* this year, but another year, try the excursions in another unit.

Tips for the Linguistically Diverse Classroom At strategic points in each unit, you will find concrete suggestions for simple modifications of the teaching strategies to encourage the participation of all students. Many of these tips offer alternative ways to elicit critical thinking from students at varying levels of English proficiency, as well as from other students who find it difficult to verbalize their thinking.

The tips are supported by suggestions for specific vocabulary work to help ensure that all students can participate fully in the investigations. The Preview for the Linguistically Diverse Classroom (p. 14) lists important words that are assumed as part of the working vocabulary of the unit. Second-language learners will need to become familiar with these words in order to understand the problems and activities they will be doing. These terms can be incorporated into students' second-language work before or during the unit. Activities that can be used to present the words are found in the appendix, Vocabulary Support for Second-Language Learners (p. 110).

In addition, ideas for making connections to students' language and cultures, included on the Preview page, help the class explore the unit's concepts from a multicultural perspective.

Session Follow-Up

Homework Homework is suggested on a regular basis in the grade 5 units. The homework may be used for (1) review and practice of work done in class; (2) preparation for activities coming up—for example, collecting data for a class project; or (3) involving and informing family members.

Some units in the *Investigations* curriculum have more homework than others, simply because it makes sense for the mathematics that's going on. Other units rely on manipulatives that most students won't have at home, making homework difficult. In any case, homework should always be directly connected to the investigations in the unit, or to work in previous units—never sheets of problems just to keep students busy.

Extensions These follow-up activities are opportunities for some or all students to explore a topic in greater depth or in a different context. They are not designed for "fast" students; mathematics is a multifaceted discipline, and different students will want to go further in different investigations. Look for and encourage the sparks of interest and enthusiasm you see in your students, and use the extensions to help them pursue these interests.

Family Letter A letter that you can send home to students' families is included with the blackline masters for each unit. We want families to be informed about the mathematics work in your classroom; they should be encouraged to participate in and support their children's work. A reminder to send home the letter appears in one of the early investigations. These letters are also available separately in Spanish, Vietnamese, Cantonese, Hmong, and Cambodian.

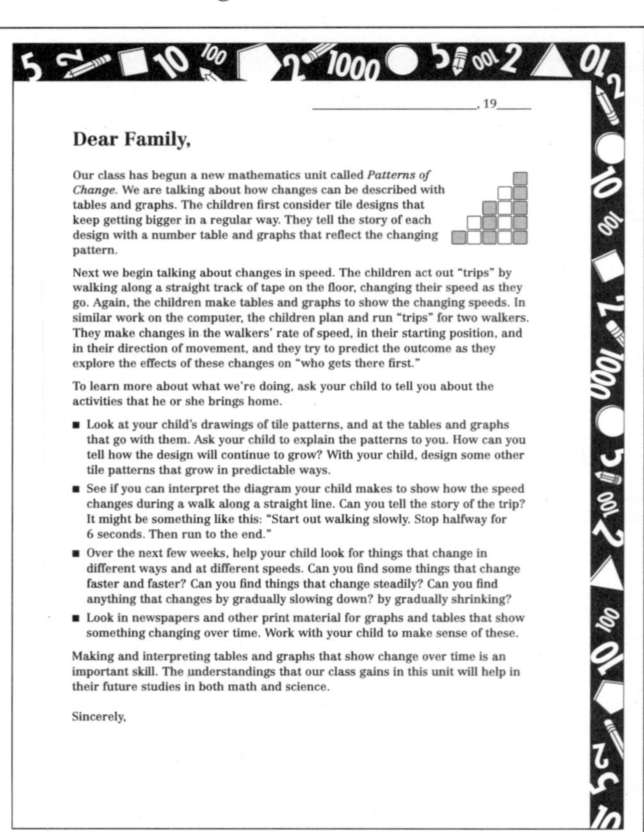

______________, 19____

Dear Family,

Our class has begun a new mathematics unit called *Patterns of Change*. We are talking about how changes can be described with tables and graphs. The children first consider tile designs that keep getting bigger in a regular way. They tell the story of each design with a number table and graphs that reflect the changing pattern.

Next we begin talking about changes in speed. The children act out "trips" by walking along a straight track of tape on the floor, changing their speed as they go. Again, the children make tables and graphs to show the changing speeds. In similar work on the computer, the children plan and run "trips" for two walkers. They make changes in the walkers' rate of speed, in their starting position, and in their direction of movement, and they try to predict the outcome as they explore the effects of these changes on "who gets there first."

To learn more about what we're doing, ask your child to tell you about the activities that he or she brings home.

- Look at your child's drawings of tile patterns, and at the tables and graphs that go with them. Ask your child to explain the patterns to you. How can you tell how the design will continue to grow? With your child, design some other tile patterns that grow in predictable ways.
- See if you can interpret the diagram your child makes to show how the speed changes during a walk along a straight line. Can you tell the story of the trip? It might be something like this: "Start out walking slowly. Stop halfway for 6 seconds. Then run to the end."
- Over the next few weeks, help your child look for things that change in different ways and at different speeds. Can you find some things that change faster and faster? Can you find things that change steadily? Can you find anything that changes by gradually slowing down? by gradually shrinking?
- Look in newspapers and other print material for graphs and tables that show something changing over time. Work with your child to make sense of these.

Making and interpreting tables and graphs that show change over time is an important skill. The understandings that our class gains in this unit will help in their future studies in both math and science.

Sincerely,

Materials

A complete list of the materials needed for the unit is found on p. 11. Some of these materials are available in a kit for the *Investigations* grade 5 curriculum. Individual items can also be purchased as needed from school supply stores and dealers.

In an active mathematics classroom, certain basic materials should be available at all times: interlocking cubes, pencils, unlined paper, graph paper, calculators, things to count with, and measuring tools. Some activities in this curriculum require scissors and glue sticks or tape. Stick-on notes and large paper are also useful materials throughout.

So that students can independently get what they need at any time, they should know where these materials are kept, how they are stored, and how they are to be returned to the storage area. For example, interlocking cubes are best stored in towers of ten; then, whatever the activity, they should be returned to storage in groups of ten at the end of the hour. You'll find that establishing such routines at the beginning of the year is well worth the time and effort.

Student Sheets and Teaching Resources

Reproducible pages to help you teach the unit are found at the end of this book. These include masters for making overhead transparencies and other teaching tools, as well as student recording sheets.

Many of the field-test teachers requested more sheets to help students record their work, and we have tried to be responsive to this need. At the same time, we think it's important that students find their own ways of organizing and recording their work. They need to learn how to explain their thinking with both drawings and written words, and how to organize their results so someone else can understand them.

To ensure that students get a chance to learn how to represent and organize their own work, we deliberately do not provide student sheets for every activity. We recommend that your students keep a mathematics notebook or folder so that their work, whether on reproducible sheets or their own paper, is always available to them for reference.

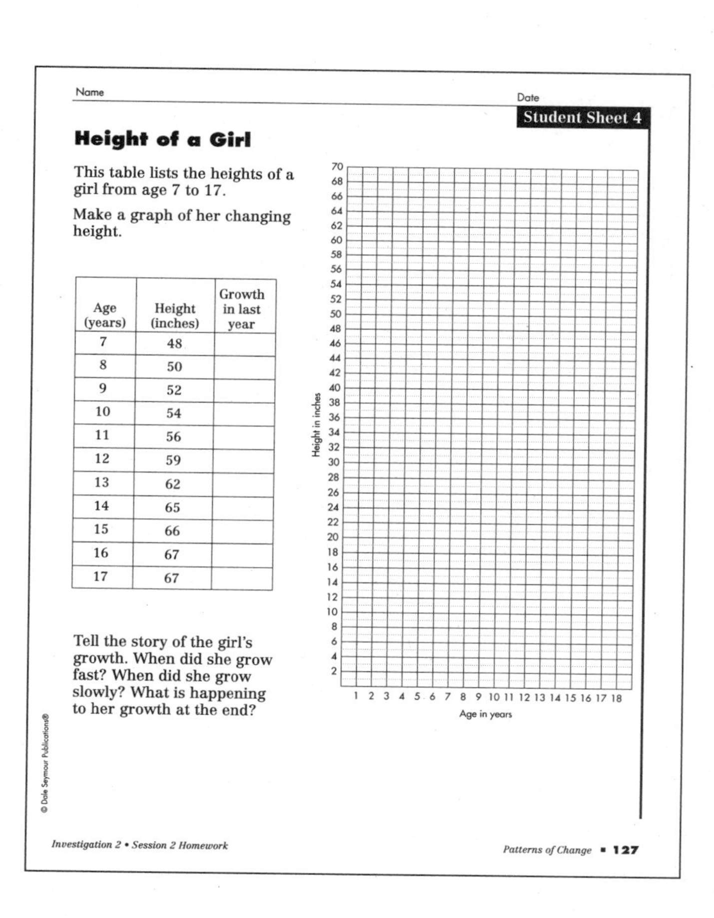

Name Date

Student Sheet 4

Height of a Girl

This table lists the heights of a girl from age 7 to 17.

Make a graph of her changing height.

Age (years)	Height (inches)	Growth in last year
7	48	
8	50	
9	52	
10	54	
11	56	
12	59	
13	62	
14	65	
15	66	
16	67	
17	67	

Tell the story of the girl's growth. When did she grow fast? When did she grow slowly? What is happening to her growth at the end?

Height in inches: 2 4 6 8 10 12 14 16 18 20 22 24 26 28 30 32 34 36 38 40 42 44 46 48 50 52 54 56 58 60 62 64 66 68 70

Age in years: 1 2 3 4 5 6 7 8 9 10 11 12 13 14 15 16 17 18

© Dale Seymour Publications®

Investigation 2 • Session 2 Homework

Patterns of Change ▪ 127

Help for You, the Teacher

Because we believe strongly that a new curriculum must help teachers think in new ways about mathematics and about their students' mathematical thinking processes, we have included a great deal of material to help you learn more about both.

About the Mathematics in This Unit This introductory section (p. 12) summarizes for you the critical information about the mathematics you will be teaching. This will be particularly valuable to teachers who are accustomed to a traditional textbook-based curriculum.

Teacher Notes These reference notes provide practical information about the mathematics you are teaching and about our experience with how students learn. Many of the notes were written in response to actual questions from teachers, or to discuss important things we saw happening in the field-test classrooms. Some teachers like to read them all before starting the unit, then review them as they come up in particular investigations.

Dialogue Boxes Sample dialogues demonstrate how students typically express their mathematical ideas, what issues and confusions arise in their thinking, and how some teachers have guided class discussions. These dialogues are based on the extensive classroom testing of this curriculum; many are word-for-word transcriptions of recorded class discussions. They are not always easy reading; sometimes it may take some effort to unravel what the students are trying to say. But this is the value of these dialogues; they offer good clues to how your students may develop and express their approaches and strategies, helping you prepare for your own class discussions.

Where to Start You may not have time to read everything the first time you use this unit. As a first-time user, you will likely focus on understanding the activities and working them out with your students. Read completely through each investigation before starting to present it.

When you next teach this same unit, you can begin to read more of the background. Each time you present this unit, you will learn more about how your students understand the mathematical ideas. The first-time user of *Patterns of Change* should read the following:

- About the Mathematics in This Unit (p. 12)
- Teacher Note: Four Patterns (p. 36)
- Teacher Note: Invented Representations of Trips (p. 46)
- Teacher Note: Tables That Show Changing Speeds (p. 54)
- Teacher Note: About the *Trips* Software (p. 78)

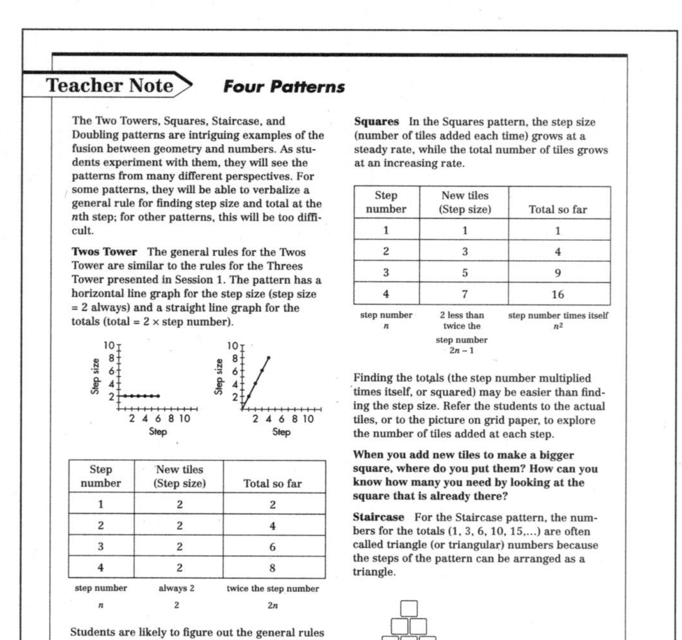

Teacher Note **Four Patterns**

The Two Towers, Squares, Staircase, and Doubling patterns are intriguing examples of the fusion between geometry and numbers. As students experiment with them, they will see the patterns from many different perspectives. For some patterns, they will be able to verbalize a general rule for finding step size and total at the *n*th step; for other patterns, this will be too difficult.

Twos Tower The general rules for the Twos Tower are similar to the rules for the Threes Tower presented in Session 1. The pattern has a horizontal line graph for the step size (step size = 2 always) and a straight line graph for the totals (total = 2 × step number).

Step number	New tiles (Step size)	Total so far
1	2	2
2	2	4
3	2	6
4	2	8
step number n	always 2 2	twice the step number $2n$

Students are likely to figure out the general rules for this pattern. You can use the totals in this

Squares In the Squares pattern, the step size (number of tiles added each time) grows at a steady rate, while the total number of tiles grows at an increasing rate.

Step number	New tiles (Step size)	Total so far
1	1	1
2	3	4
3	5	9
4	7	16
step number n	2 less than twice the step number $2n - 1$	step number times itself n^2

Finding the totals (the step number multiplied times itself, or squared) may be easier than finding the step size. Refer the students to the actual tiles, or to the picture on grid paper, to explore the number of tiles added at each step.

When you add new tiles to make a bigger square, where do you put them? How can you know how many you need by looking at the square that is already there?

Staircase For the Staircase pattern, the numbers for the totals (1, 3, 6, 10, 15,...) are often called triangle (or triangular) numbers because the steps of the pattern can be arranged as a triangle.

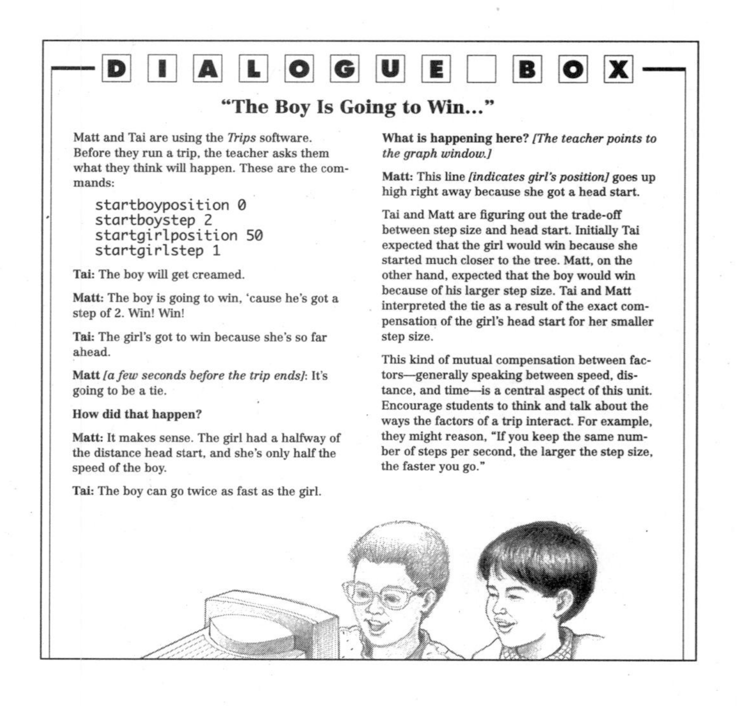

DIALOGUE BOX

"The Boy Is Going to Win..."

Matt and Tai are using the *Trips* software. Before they run a trip, the teacher asks them what they think will happen. These are the commands:

```
startboyposition 0
startboystep 2
startgirlposition 50
startgirlstep 1
```

Tai: The boy will get creamed.

Matt: The boy is going to win, 'cause he's got a step of 2. Win! Win!

Tai: The girl's got to win because she's so far ahead.

Matt *[a few seconds before the trip ends]*: It's going to be a tie.

How did that happen?

Matt: It makes sense. The girl had a halfway of the distance head start, and she's only half the speed of the boy.

Tai: The boy can go twice as fast as the girl.

What is happening here? *[The teacher points to the graph window.]*

Matt: This line *[indicates girl's position]* goes up high right away because she got a head start.

Tai and Matt are figuring out the trade-off between step size and head start. Initially Tai expected that the girl would win because she started much closer to the tree. Matt, on the other hand, expected that the boy would win because of his larger step size. Tai and Matt interpreted the tie as a result of the exact compensation of the girl's head start for her smaller step size.

This kind of mutual compensation between factors—generally speaking between speed, distance, and time—is a central aspect of this unit. Encourage students to think and talk about the ways the factors of a trip interact. For example, they might reason, "If you keep the same number of steps per second, the larger the step size, the faster you go."

Teacher Checkpoints As a teacher of the *Investigations* curriculum, you observe students daily, listen to their discussions, look carefully at their work, and use this information to guide your teaching. We have designated Teacher Checkpoints as natural times to get an overall sense of how your class is doing in the unit.

The Teacher Checkpoints provide a time for you to pause and reflect on your teaching plan while observing students at work in an activity. These sections offer tips on what you should be looking for and how you might adjust your pacing. Are most students fluent with strategies for solving a particular kind of problem? Are they just starting to formulate good strategies? Or are they still struggling with how to start?

Depending on what you see as the students work, you may want to spend more time on similar problems, change some of the problems to use smaller numbers, move quickly to more challenging material, modify subsequent activities for some students, work on particular ideas with a small group, or pair students who have good strategies with those who are having more difficulty.

In *Patterns of Change* you will find three Teacher Checkpoints:

Investigation 1, Session 1:
Designing a Growing Tile Pattern (p. 22)

Investigation 2, Session 3:
Making Tables for Stories (p. 57)

Investigation 3, Session 2:
Running and Recording Trips (p. 84)

Embedded Assessment Activities Use the built-in assessments included in this unit to help you examine the work of individual students, figure out what it means, and provide feedback. From the students' point of view, the activities you will be using for assessment are no different from any others; they don't look or feel like traditional tests.

These activities sometimes involve writing and reflecting, at other times a brief interaction between student and teacher, and in still other instances the creation and explanation of a product.

In *Patterns of Change* you will find two assessment activities:

Investigation 2, Session 5:
Matching Stories, Tables, and Graphs (p. 66)

Investigation 3, Sessions 5-6:
Graphs: What Story Do They Tell? (p. 99)

Teachers find the hardest part of the assessment to be interpreting their students' work. If you have used a process approach to teaching writing, you will find our mathematics approach familiar. To help with interpretation, we provide guidelines and questions to ask about the students' work. In some cases we include a Teacher Note with specific examples of student work and a commentary on what it indicates. This framework can help you determine how your students are progressing.

As you evaluate students' work, it's important to remember that you're looking for much more than the "right answer." You'll want to know what their strategies are for solving the problem, how well these strategies work, whether they can keep track of and logically organize an approach to the problem, and how they make use of representations and tools to solve the problem.

Ongoing Assessment Good assessment of student work involves a combination of approaches. Some of the things you might do on an ongoing basis include the following:

- **Observation** Circulate around the room to observe students as they work. Watch for the development of their mathematical strategies, and listen to their discussions of mathematical ideas.
- **Portfolios** Ask students to document their work, in journals, notebooks, or portfolios. Periodically review this work to see how their mathematical thinking and writing are changing. Some teachers have students keep a notebook or folder for each unit, while others prefer one mathematics notebook or a portfolio of selected work for the entire year. Take time at the end of each unit for students to choose work for their portfolios. You might also have them write about what they've learned in the unit.

Patterns of Change

OVERVIEW

Content of This Unit Using plastic tiles, students experiment with forms of geometric growth that express number patterns. Students show their growing patterns in number tables and on graphs, and distinguish between growth, shrinkage, and oscillation, as well as between steady and accelerated growth.

Students then use number tables, graphs, and written "motion stories" to describe walks along a straight line or "track," noting the patterns that show increasing or decreasing speed. In the *Trips* software, which simulates a boy and a girl walking along two tracks, and in parallel activities off computer, students explore the trade-off between factors that influence the outcome of a trip, including step size (which, at 1 step per second, represents speed) and starting position. The unit culminates with the analysis of the relationship between graphs of position vs. time and graphs of step size vs. time.

Connections with Other Units If you are doing the full-year *Investigations* curriculum in the suggested sequence for grade 5, this is the seventh of nine units. It builds on ideas developed in the grade 3 unit *Up and Down the Number Line* and the grade 4 unit *Changes Over Time.* If your students have not made line graphs to show something changing over time, such as plant growth, you might use the grade 4 unit before starting *Patterns of Change.*

This unit can also be used successfully at grade 6, depending on the previous experience and needs of your students.

Investigations Curriculum ■ Suggested Grade 5 Sequence

Mathematical Thinking at Grade 5 (Introduction and Landmarks in the Number System)

Picturing Polygons (2-D Geometry)

Name That Portion (Fractions, Percents, and Decimals)

Between Never and Always (Probability)

Building on Numbers You Know (Computation and Estimation Strategies)

Measurement Benchmarks (Estimating and Measuring)

▶ *Patterns of Change* (Tables and Graphs)

Containers and Cubes (3-D Geometry: Volume)

Data: Kids, Cats, and Ads (Statistics)

Investigation 1 • Number Patterns in Changing Shapes

Class Sessions	Activities	Pacing	Ten-Minute Math
Sessions 1 and 2 (p. 18) ANALYZING TILE PATTERNS	Exploring Tile Patterns Other Patterns and Their Graphs ■ Teacher Checkpoint: Designing a Growing Tile Pattern ■ Homework	2 hrs	Nearest Answer
Sessions 3 and 4 (p. 28) FOUR GROWING TILE PATTERNS	Making Tables and Graphs for Tile Patterns Comparing Graph Shapes Investigating a Pattern in Depth Presenting the Posters ■ Homework ■ Extension	2 hrs	

Investigation 2 • Motion Stories, Graphs, and Tables

Class Sessions	Activities	Pacing	Ten-Minute Math
Session 1 (p. 42) DESCRIBING CHANGING SPEEDS	Describing Fast and Slow Trips Representing Changes of Speed ■ Homework	1 hr	Nearest Answer
Session 2 (p. 47) FROM BEANBAGS TO TABLES	Comparing Fast and Slow Trips Marking and Guessing Trips Making Tables to Show Trips ■ Homework	1 hr	
Session 3 (p. 55) TABLES FOR STORIES	Interpreting the Table of Heights ■ Teacher Checkpoint: Making Tables for Stories Matching Tables to Stories Similarities Among Tables ■ Homework	1 hr	
Session 4 (p. 62) GRAPHS FOR TABLES	Homework Review: Graph of a Trip Making Graphs from Tables Interpreting Graphs ■ Homework ■ Extension	1 hr	
Session 5 (p. 66) STORIES, TABLES, AND GRAPHS	■ Assessment: Matching Stories, Tables, and Graphs ■ Homework	1 hr	

Continued on next page

Investigation 3 • Computer Trips on Two Tracks

Class Sessions	Activities	Pacing	Ten-Minute Math
Session 1 (p. 71) WAYS OF MAKING TRIPS	Showing Speed with Step Size Trips Along a Meterstick Trips on the Computer Exploring the *Trips* Software	1 hr	Graph Stories
Session 2 (p. 81) TRIPS ON TWO TRACKS	Working with Motion Stories ■ Teacher Checkpoint: Running and Recording Trips ■ Homework ■ Extension	1 hr	
Session 3 (p. 87) DIFFERENT KINDS OF TRIPS	How Settings 2 and 3 Work Running Trips in Setting 2 and Setting 3 ■ Homework ■ Extension	1 hr	
Session 4 (p. 93) MORE MATCH-UPS	Comparing Tables and Graphs ■ Extension	1 hr	
Sessions 5 and 6 (p. 95) TWO TYPES OF GRAPHS	Showing a Walk with Two Graphs Graphing and Guessing Mystery Walks ■ Assessment: Graphs: What Story Do They Tell? ■ Homework ■ Extension	2 hrs	
Session 7 (Excursion)* (p. 102) ANIMATION	Animated Flip Books Flip Books with Two Changes Choosing Student Work to Save ■ Homework	1 hr	

* Excursions can be omitted without harming the integrity or continuity of the unit, but offer good mathematical work of you have time to include them.

MATERIALS LIST

Following are the basic materials needed for the activities in this unit. Items marked with an asterisk are provided with the *Investigations* Materials Kit for grade 5.

* * Square tiles in mixed colors: 1 bucket of 400 per group of 6–8 students
* Overhead projector
* Overhead pens, in colors of tiles (optional)
* Transparent tiles for overhead: 10 each of four colors (optional)
* * Metersticks: 1 per 8–12 students
* Stopwatches or wall clock that shows seconds: 1 per 4–6 students
* Small beanbags in two colors (see p. 41 for ideas for making these): 12 per 4–6 students
* Shallow containers to hold 12 beanbags: 1 per 4–6 students
* Cuisenaire® rods: 3–4 sets of 10. If you do not have these, substitute centimeter cubes taped together in varying lengths or narrow strips of tagboard cut in 1- to 10-centimeter lengths.
* Computers—Macintosh II or above, with 4 MB of internal memory (RAM) and Apple System Software 7.0 or later: enough for half the class, working in pairs, to use at once
* * Apple Macintosh disk, *Trips*™ (provided with this unit)
* Stick-on notes: 2 pads for the class to share; 1 pad per student of 2-by-3-inch size (for making flip books)
* Wide (2-inch) masking tape: about 50 meters
* Adding machine tape: 5–6 meters
* Colored markers or crayons for every student
* Scissors: 1 per pair
* Tape or glue sticks: 1 per pair
* Large paper for posters: 1 sheet per 3–4 students
* Chart paper
* Calculators: always available

The following materials are provided at the end of this unit as blackline masters.

Family Letter (p. 120)

Student Sheets 1–19 (starting p. 121)

Teaching Resources:

- One-Centimeter Graph Paper (p. 151)
- Graph Shapes (for Ten-Minute Math) (p. 152)

Things change over time—sometimes steadily, sometimes at varying rates. Figuring out how quickly something grows or declines is essential not just in higher mathematics, but in the sciences and social sciences as well. Elementary students need some experience in describing, representing, and comparing *rates of change;* that is the focus of this unit. Following are some of the major mathematical ideas that students encounter:

Understanding the relationship between rate of change and accumulated change. An appreciation of the connections between rate of change and accumulated change is a foundational skill for students as they proceed through their study of mathematics toward calculus. Students explore these ideas about that relationship:

- A constant positive rate of change corresponds to a steady increase in the accumulated changes. In terms of the "trips" students explore, moving at a constant speed implies that the distance traveled increases steadily.
- We can accumulate the same total change with many small successive changes or a few big ones. In "trips" terms, we can travel the same distance in the same time with many small steps or a few big steps.

These ideas have profound mathematical implications that can be expressed in the shape of graphs, in number patterns, and in geometric growth.

Using the language of motion to describe mathematical behavior. Motion, with its pervasive presence in daily life, offers powerful metaphors and language to use in considering symbolic expressions. For example, consider two number series:

1, 3, 5, 7, ...
1, 2, 4, 8, 16, ...

We can compare them by saying that the first increases at a steady rate and the second grows faster and faster. Even though nothing is actually moving, the language of motion *(steady, faster and faster)* allows us to communicate the change and the overall trends in these number patterns.

A crucial part of mathematical understanding is being able to look at symbolic expressions and "see" in these static forms the dynamics of events changing over time.

Seeing connections between graphs, number tables, and stories. To make sense of a mathematical model, we need to be able to tell stories that express events as symbolic features. For example, we can show one person catching up to another (an event) as two graphs crossing each other (a symbolic feature).

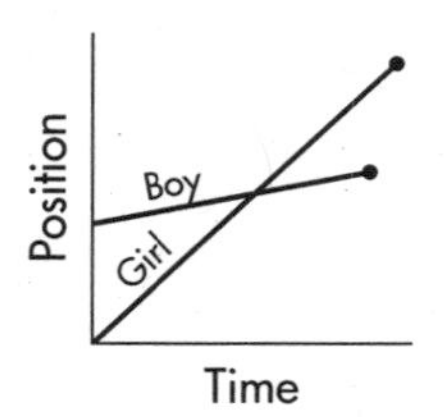

We can show climbing a particular staircase (an event) as numbers in a table increasing by 2's (a symbolic feature).

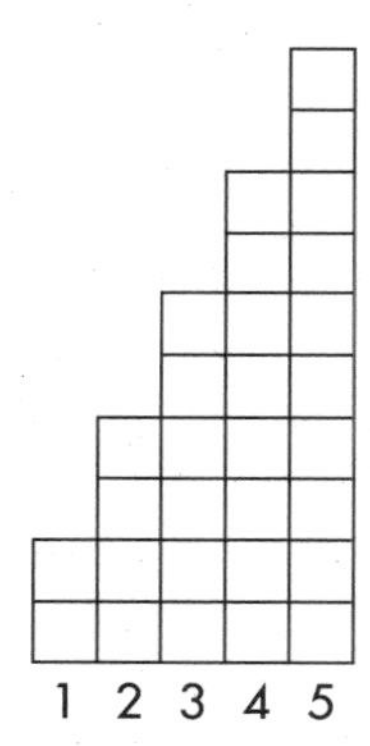

Step	Total
1	2
2	4
3	6
4	8
5	10

Students experiment with the notion that the same story can be told in many mathematical forms, such as graphs, patterns of marks along a line, and number tables, which are equivalent precisely because they tell the same story.

Distinguishing across-time and simultaneous relationships. In observing something that changes, we can take a *diachronic* (across-time) perspective or a *synchronic* (simultaneous) perspective. For example, we can study someone's growth across the years, perhaps by looking at the person's height in each past year. Or, we can compare someone's height to the height of others at a given time, perhaps by arranging the names of a group of people according to their height.

In symbolic terms, we can describe the number table below by noticing that N increases by 1's and P by 2's (a diachronic view). Or, we can point out that P is always $2 \times N$ (a synchronic view).

N	P
1	2
2	4
3	6
4	8
5	10

Fluency in moving back and forth between these two perspectives is central to the mathematical analysis of change.

Mathematical Emphasis At the beginning of each investigation, the Mathematical Emphasis section tells you what is most important for students to learn about during that investigation. Many of these understandings and processes are difficult and complex. Students gradually learn more and more about each idea over many years of schooling. Individual students will begin and end the unit with different levels of knowledge and skill, but all will become more able to make, interpret, and compare graphs and tables that show something changing over time.

In the *Investigations* curriculum, mathematical vocabulary is introduced naturally during the activities. We don't ask students to learn definitions of new terms; rather, they come to understand such words as *factor* or *area* or *symmetry* by hearing them used frequently in discussion as they investigate new concepts. This approach is compatible with current theories of second-language acquisition, which emphasize the use of new vocabulary in meaningful contexts while students are actively involved with objects, pictures, and physical movement.

Listed below are some key words used in this unit that will not be new to most English speakers at this age level, but may be unfamiliar to students with limited English proficiency. You will want to spend additional time working on these words with your students who are learning English. If your students are working with a second-language teacher, you might enlist your colleague's aid in familiarizing students with these words, before and during this unit. In the classroom, look for opportunities for students to hear and use these words. Activities you can use to present the words are given in the appendix, Vocabulary Support for Second-Language Learners (p. 110).

grow, shrink Students note whether the regular patterns they are studying are growing, shrinking, or doing both in turns.

position, start, end As students represent trips in various ways, they discuss the start and end positions of people or markers as they move along a path.

change, speed, faster, slower, increase, steady Throughout the unit, students note the way in which patterns or trips change, and whether the rate of change is getting faster or slower or remaining steady.

steep, flat, slope Students graph the patterns they discover, noting when the patterns are growing or shrinking rapidly, resulting in a steep slope, and when they are not changing, resulting in a flat horizontal line.

Multicultural Extensions for All Students

Whenever possible, encourage students to share words, objects, customs, or any aspects of daily life from their own cultures and backgrounds that are relevant to the activities in this unit. For example, students might bring in tables and graphs from foreign-language magazines and newspapers and work together as a class to attempt to interpret them.

Investigations

INVESTIGATION 1

Number Patterns in Changing Shapes

What Happens

Sessions 1 and 2: Analyzing Tile Patterns
Students explore tile patterns and analyze how they "grow," deciding how to extend each pattern. In a table and on graphs, they show the number of new tiles added at each step and the total number of tiles used in the pattern so far. They create a tile pattern of their own and make a table and graphs that reflect the pattern.

Sessions 3 and 4: Four Growing Tile Patterns
Students analyze four tile patterns with these growing totals: 2, 4, 6, 8,...; 1, 4, 9, 16,...; 1, 3, 6, 10,...; and 2, 6, 14, 30,.... In tables and on graphs, students show how the tile patterns and the number series grow (whether at a steady or accelerating rate). Small groups investigate one of the four patterns in depth. They make a poster that shows the tile pattern shapes, a table, two graphs, and a written description of how their pattern grows.

Mathematical Emphasis

- Building designs that change in a regular way
- Building designs that grow according to number patterns
- Predicting later steps of number patterns and designs
- Making tables and graphs to display number patterns
- Comparing sequences of numbers of new tiles and numbers of accumulated tiles
- Using the language of speed and motion to describe number patterns

What to Plan Ahead of Time

Materials

- Square tiles in mixed colors: 1 bucket of 400 per 6–8 students (all sessions)
- Markers, pencils, or crayons in the colors of the tiles (all sessions)
- Overhead projector (Sessions 1–3)
- Pens for overhead projector, in colors of tiles (Session 1, optional)
- Transparent tiles for overhead projector: 10 each of four colors (Sessions 1–3, optional)
- Large paper for posters: 1 sheet per group of 3–4 students (Sessions 3–4)
- Scissors and tape for each group (Sessions 3–4)

Other Preparation

- Duplicate student sheets and teaching resources (located at the end of this unit) as follows:

  For Sessions 1 and 2

  Student Sheet 1, Tile Pattern Template (p. 121): 2–3 per student and 2 transparencies

  Centimeter graph paper (p. 151): 1 sheet per pair and 1 per student (homework)

  Family letter (p. 120): 1 per student. Remember to sign it before copying.

  For Sessions 3 and 4

  Student Sheet 2, Growing Tile Patterns (pp. 122–125): 1 per student, and transparencies of first two patterns (Twos Tower and Squares)

  Student Sheet 1, Tile Pattern Template (p. 121): 1 per student (homework)

  Centimeter graph paper (p. 151): 2 sheets per 3–4 students, plus 1 per student (homework)

Sessions 1 and 2

Analyzing Tile Patterns

Materials

- Overhead projector
- Transparent tiles for overhead (optional)
- Transparencies of Student Sheet 1
- Overhead pens in colors of tiles (optional)
- Centimeter graph paper (1 per pair and 1 per student, homework)
- Square tiles in mixed colors (1 bucket of 400 per 6–8 students)
- Student Sheet 1 (2–3 per student)
- Markers, pencils, or crayons in the colors of the tiles (optional)
- Family letter (1 per student)

What Happens

Students explore tile patterns and analyze how they "grow," deciding how to extend each pattern. In a table and on graphs, they show the number of new tiles added at each step and the total number of tiles used in the pattern so far. They create a tile pattern of their own and make a table and graphs that reflect the pattern. Their work focuses on:

- predicting later steps in tile patterns and in number patterns
- building designs that change in a regular way
- making tables and graphs that describe growing tile patterns
- finding patterns in step numbers and totals

Activity

Exploring Tile Patterns

To introduce the idea of making a pattern that grows in a regular way, use transparent tiles on an overhead projector or make tile designs on a flat surface where students can gather around. Build the beginning of a Threes Tower—a growing stack of rows of three tiles. Start with one color at the bottom and build up, changing the color for each new row. (Throughout this investigation, we will illustrate patterns with tiles in four colors, red (R), blue (B), green (G), and yellow (Y), in no particular order.)

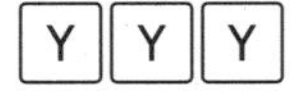

As you make the first three rows, talk about what you are doing.

I'm building a pattern. For each new step of the pattern, I'll change the tile color so the step is clear.

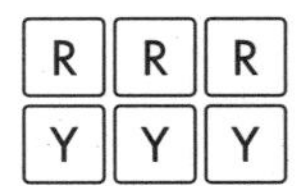

Explain that *where* the new tiles are added is important, but the order of the colors is not.

B	B	B
R	R	R
Y	Y	Y

What do you think my next step will be? How many new tiles will I add?

Take suggestions about how many tiles to add and where to put them. When students have the idea of how to continue this pattern, start a three-column table on a blank overhead transparency or on the board, showing the step number, the number of new tiles added at each step, and the total number of tiles used thus far. With students' suggestions, continue the table a few rows.

Step number	New tiles (step size)	Total so far
1	3	3
2	3	6
3	3	9
4		
5		

Making a table is a way of seeing if there is a number pattern in the way this design is growing. What patterns do you see? As we continue adding tiles to this pattern, what numbers will we put in the column for new tiles? We will call this number the *step size*. What numbers will we put in the column for total tiles used so far?

To encourage students to read across rows in the table, ask questions about some steps you have not reached yet and perhaps about some steps that are beyond the table. You might cover the row just above the one you are asking about so that students can't build onto it to get the answer.

What will the *number of new tiles* be for the fifth step? for the seventh step? for any step?

What will the *total number of tiles* be for the fifth step? for the seventh step? If I kept on building, what would be the total after the tenth step? the twentieth step? If I tell you the step number, what calculations would you do to get the total?

When students identify the patterns in the step size and total, add notes at the bottom of the table columns to describe the patterns.

Step number	New tiles (step size)	Total so far
1	3	3
2	3	6
3	3	9
4	3	12
5	3	15
6	3	18
7	3	21
8	3	24
	always 3	Multipy step number x 3

When you have completed the table through step 8, ask about the overall pattern.

How is this pattern growing? Is the step size growing? (No, it stays constant at 3.) **Is the total growing? Is it growing steadily, by the same amount every step, or is it growing faster and faster?** (The total grows steadily, adding 3 more for each step.)

Activity

Other Patterns and Their Graphs

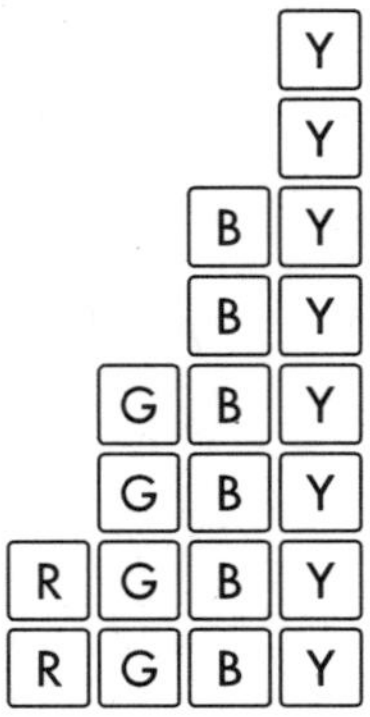

			Y
			Y
		B	Y
		B	Y
	G	B	Y
	G	B	Y
R	G	B	Y
R	G	B	Y

Start another tile design, as shown below (at left). We call this one the Double-Step Staircase, but students may see it differently and want to call it by another name. Use whatever name they are comfortable with.

Describe each step as you build:

In the first step I'm placing two reds. In the second step I'm placing four greens. For the third step I'm taking six blues. How do you think I should place the blues?

What do you think the fourth step will be? How many yellows should I use? How should I place them?

On a transparency of Student Sheet 1, Tile Pattern Template, draw the Double-Step Staircase tile design. Ask students to describe how the general shape of the design is growing as you add tiles. Is it getting wider? taller? In what direction is it growing? Some students may observe that the design is growing taller at a faster rate than it is growing wider; some may even notice that it is growing taller exactly twice as fast as it is growing wider (that is, it grows 2 tiles taller for every 1 tile that it grows wider).

Point out the table below the pattern grid on Student Sheet 1. With students' suggestions, fill in the table with numbers that describe the pattern.

Start two more tile pattern designs. As necessary, remind them that each color change represents a new step.

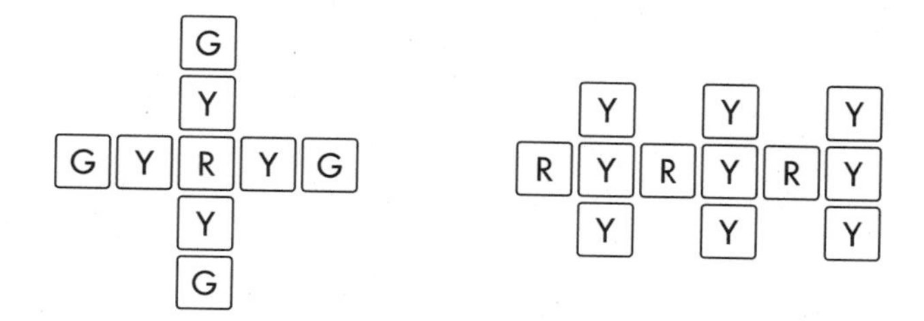

Again, encourage students to describe generally how the patterns are growing. (The pattern on the left is growing equally taller and wider in all directions at once. The pattern on the right is growing longer but not taller.)

After students have established how these tile patterns are to be continued, hand out a sheet of centimeter graph paper to each pair. Pairs copy one of the two patterns and continue it on the graph paper, then make a table to go with it.

When everyone has copied a design and begun the table, return to the overhead to demonstrate how patterns like these can be represented on graphs.

Show again your transparency of the Double-Step Staircase design. Ask students to read numbers from the table as you draw the corresponding graphs: first the smaller graph showing how the number of new tiles (step size) changes, then the graph on the right, showing how the total number of tiles grows. As students read each column, draw dots where graph lines intersect, showing (on the step-size graph at left) the number of new tiles used at each step (2, 4, 6, 8, 10...), and (on the totals graph at right) the total number of tiles in the pattern after each step (2, 6, 12, 20...).

If I continued the graph of the total number of tiles, on and on, how would it look? How does it compare with the graph of step size? Can you explain why these two graphs are different?

See the **Teacher Note,** The Relationship of Step Size to Total, for a discussion of this issue.

How do the lines in these graphs show that the steps grow steadily, but the total grows faster and faster? (The step size graph is a straight line, but the total graph curves, getting steeper and steeper.)

Associating shapes of graphs with the story of growth for a pattern can be difficult; some students may need a lot of experience before the associations are clear. Students who have graphed plant growth in the grade 4 *Investigations* unit *Changes Over Time* will have made these kinds of connections.

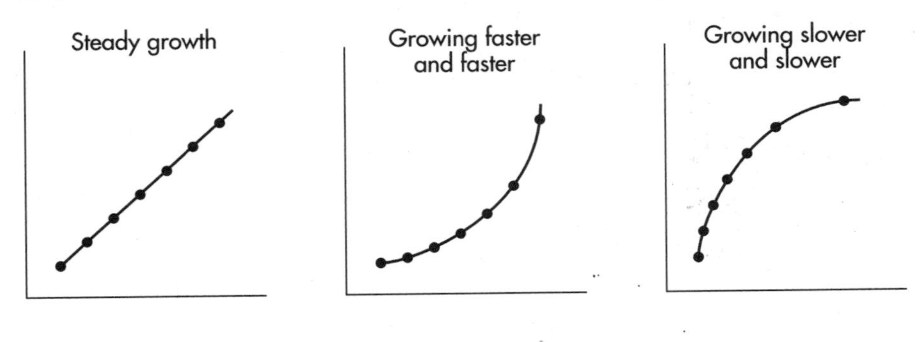

Activity

Teacher Checkpoint

Designing a Growing Tile Pattern

Distribute the buckets of tiles. Explain that students will now be making their own designs that change in a regular way, so that someone else could continue the pattern just by looking at it.

Work in pairs to make up your own growing pattern of tiles. After you do about three steps, ask other students near you to guess how tiles should be added to your design for the next few steps. Your pattern should be regular enough so that they can predict how it will grow.

As students work, check their patterns to see if you can understand where to place the next tiles. As necessary, remind them to change colors at each new step. When a pair of students have a design that grows in a regular way, give them a copy of Student Sheet 1 to record it. They draw their design on the grid, then work together to fill out the table and the two graphs. Suggest that they check their table with another pair before starting to make the graphs (if the table has errors, the graphs will, too).

Use this activity as a checkpoint to see if students are developing the idea of making a table and graphs based on a growing tile design. Watch for students who fill out their table by continuing a number pattern they think they see in the table but that does not reflect the design. Help them refer back to the design to keep the table accurate.

Look for the following as students work:

- Can students derive step size from the design?
- Can they derive the total directly from the design and also from adding the new step to the previous total?
- Can they make both graphs from the entries in their table?
- Can they interpret the shape of a graph as showing no growth, steady growth, accelerating growth, or alternating fast and slow growth?

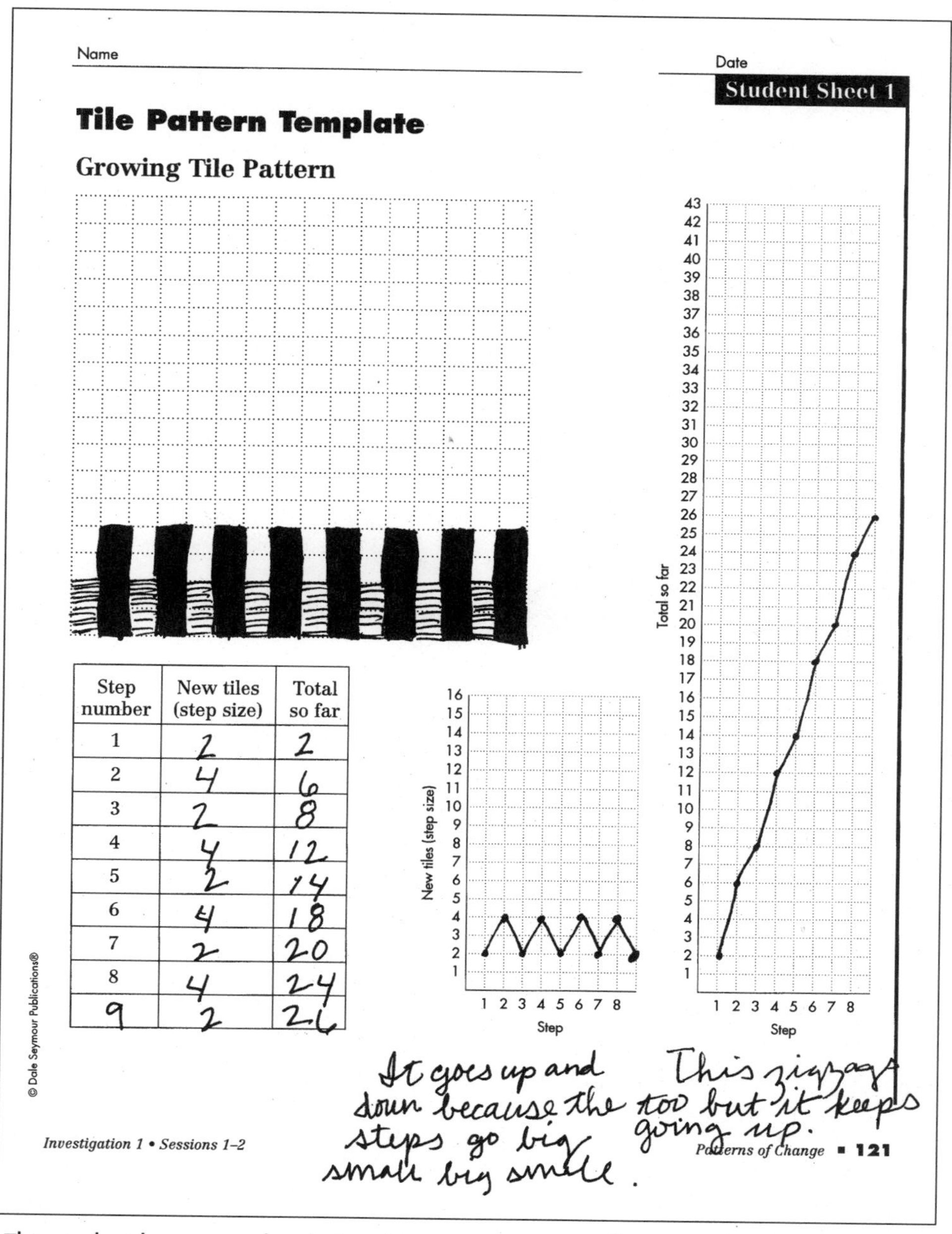

Name

Date

Student Sheet 1

Tile Pattern Template

Growing Tile Pattern

Step number	New tiles (step size)	Total so far
1	2	2
2	4	6
3	2	8
4	4	12
5	2	14
6	4	18
7	2	20
8	4	24
9	2	26

It goes up and down because the steps go big small big smlle.

This zigzags too but it keeps going up.

Investigation 1 • Sessions 1–2

Patterns of Change ■ **121**

This student has created a design that grows in an oscillating pattern.

During the next two sessions, check on students who are having difficulty with the tables and graphs. Students will be making and analyzing tables and graphs throughout this unit, so it is worth taking time in this first investigation to give them a solid footing.

Before the end of class, ask students working near each other to look at each other's designs and decide how they are changing.

Does the step size grow, or stay the same, or change in some other way? Does the total grow steadily, or faster and faster, or change in some other way?

Students write a general description of the growth patterns of their tile design—both for the steps and for the overall design.

❖ **Tip for the Linguistically Diverse Classroom** Students may describe the growth patterns orally to you, pointing to the elements they have recorded on Student Sheet 1.

If students feel ready, they might write general rules for the step size and the total. See the **Teacher Note,** Finding General Rules for Step Size and Total (p. 26), for more information.

Sessions 1 and 2 Follow-Up

Homework

- After Session 1, send home the family letter. Students also take home a piece of centimeter graph paper and draw another pattern that changes in a regular way. They can color the patterns if they wish. They also make a table showing the step size—or the number of new "tiles" (grid squares) they added to the pattern at each step—and the total number of tiles (squares) in the pattern after each step. They also describe generally how their tile pattern grows.
- After Session 2, students take home their work on tile patterns along with another copy of Student Sheet 1, Tile Pattern Template. They prepare a sheet that can be displayed, drawing one tile pattern and filling in the table and graphs. They may use any of the tile patterns discussed in class or one of their own.

The Relationship of Step Size to Total

Teacher Note

As students investigate tile designs, they identify patterns as designs that change with a predictable regularity—that is, there is a pattern to the size of the steps. We can call the number of new tiles added at each subsequent step the *step size,* and the number of tiles in the whole design after any step the *total.*

The use of these terms here prepares students for the terminology they will be using as they graph motion and varying speed in Investigations 2 and 3. The *step size* for the tile patterns (how many tiles we add in a step) is analogous to the *step size* for a "trip" along a straight track (the distance traveled in one unit of time). The total number of tiles is analogous to the position reached by the traveler, or the total distance traveled.

For any tile pattern, the step size and the total are related but grow differently from one another. In the tile design we called the Threes Tower, the *step size* does not grow at all. It stays constant (the graph is flat), so the *total* grows at a steady rate (the graph of the total is a straight line with slope 3).

Threes Tower
tile pattern

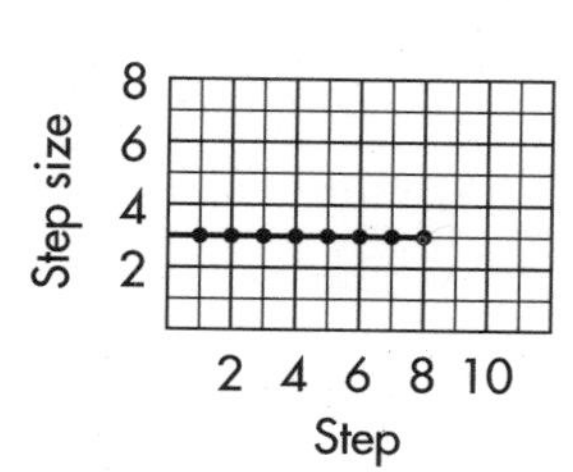

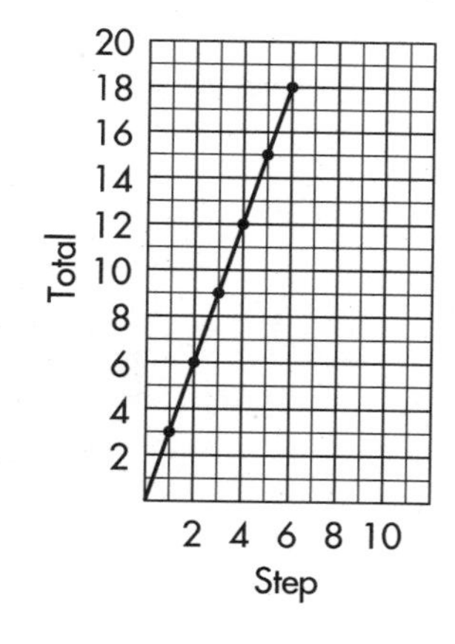

In the Double-Step Staircase design, the step size grows at a steady rate, increasing by 2 each time (the graph is a straight line with slope 2). Because each step is 2 larger than the step before, the total grows faster and faster (the graph is a curve with a steeper and steeper slope).

Double-step
Staircase
tile pattern

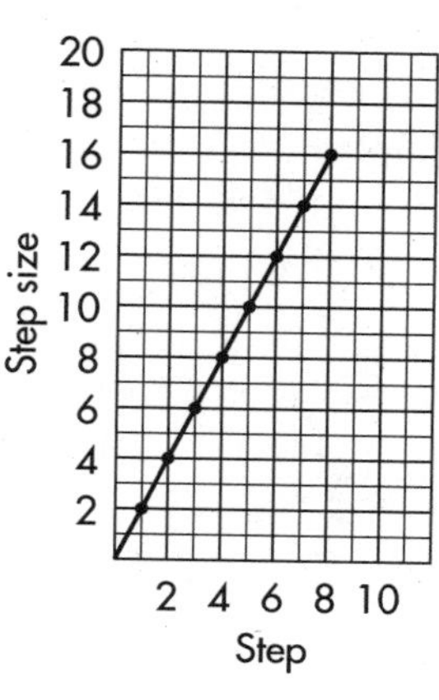

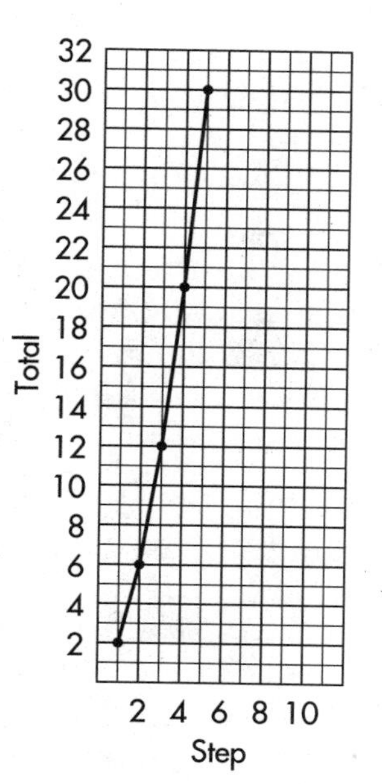

It is possible to have a decreasing step size while the total continues to increase. If we use fewer and fewer tiles each time, the step size decreases, but the total number of tiles continues to increase—although by less and less each time. If the step size continues to decrease, it will eventually reach zero and fall below zero, making the total growth stop and then decrease.

For example, imagine that we used this decreasing sequence of number of tiles:

4, 3, 2, 1, 0, –1, –2,...

At first the total number of tiles would increase by 4, 3, 2, and 1; then it would stop, having an increase of 0; and finally it would decrease, because using –1, –2,... would mean that instead of adding tiles to the pattern, we would be taking tiles away.

Teacher Note

Finding General Rules for Step Size and Total

As students make tables to reflect their growing tile designs, they look for patterns in the numbers that enable them to predict what the step size and the total for later steps will be. In their later study of mathematics, students will be expected to find a general rule that will work for any step number of a series (often called n for *number*) without finding all the terms in between. Many fifth grade students may be able to do this already for some patterns.

For example, students at this level may find general rules in the Threes Tower for both the step size (always 3) and the total ($3 \times$ step number), and in the Double-Step Staircase for step size ($2 \times$ step number). Encourage them to write the general rules they find in any informal way that makes sense. Do not push them to use n at this time.

When students are looking for general patterns in the tile designs, they will get different information from each of the representations: the design itself, the number table, and the graphs. The graph of totals gives an overview of the growth, showing *increase* or *decrease* or *oscillation.* It displays steady growth with a straight line and accelerating growth with a curved line.

The number table can be analyzed to see how early numbers can be used to find later ones. Students are likely to recognize patterns within columns first. For example, they may see in the table that the steps in the Threes Tower are always 3, and that for the total, we count by 3's. Later they may see that each total is made by adding the new step size to the previous total. Last of all, they may figure out how the numbers within one row are related—that we can get the total for the Threes Tower by multiplying the step number by 3, or we can get the step size in the Squares pattern (Student Sheet 2, page 1) by doubling the step number and subtracting 1.

To find a general pattern, it is often most helpful to look at the design itself. For example, to make the next larger step in the Squares pattern, we add a row of tiles along the top, a column along the side, and one additional tile in the corner; this is doubling the previous step number and adding 1. We might get a slightly different rule from the table: double the current step number and subtract 1. Both rules are equally descriptive of relationships in the pattern.

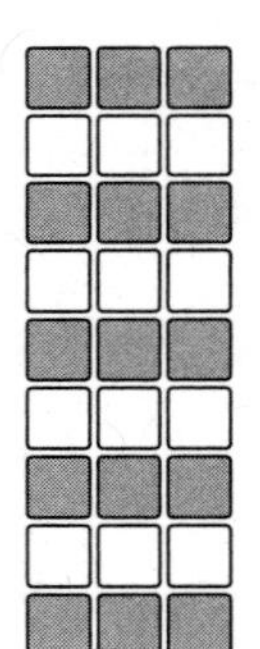

Step number	New tiles (step size)	Total so far
1	3	3
2	3	6
3	3	9
4	3	12
5	3	15
6	3	18
7	3	21
8	3	24
9	3	27

Threes Tower tile pattern

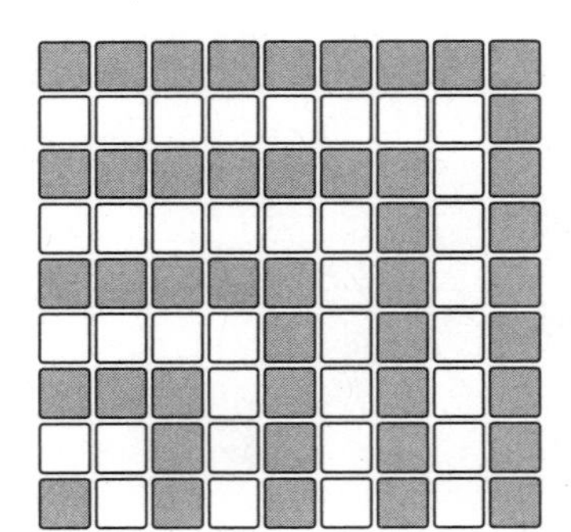

Step number	New tiles (step size)	Total so far
1	1	1
2	3	4
3	5	9
4	7	16
5	9	25
6	11	36
7	13	49
8	15	64
9	17	81

Square tile pattern

Continued on next page

Teacher Note

continued

Finding the general rule for the total number of tiles is difficult for the Double-Step Staircase pattern and for the Staircase pattern on Student Sheet 2. Avoid posing this as a question for the whole class or you will be off on a detour that may prove frustrating for many students. However, some students may be interested in these patterns; in one class, one pair worked for 20 minutes on the rule for the Double-Step Staircase total instead of going on to make a pattern of their own. They came to the conclusion (from analyzing the table) that you multiply each step number by the next step number to get the total.

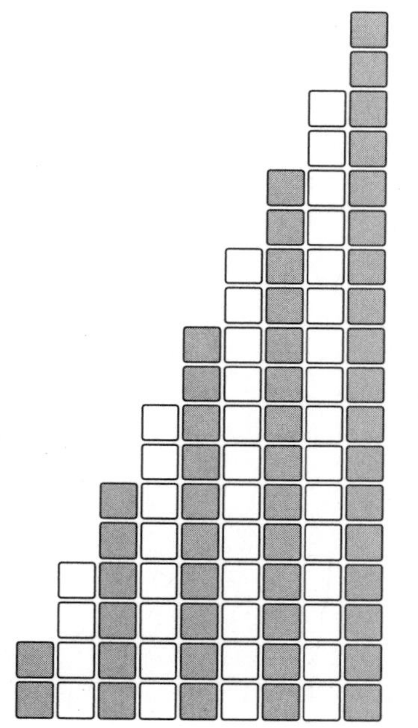

Step number	New tiles (step size)	Total so far
1	2	2
2	4	6
3	6	12
4	8	20
5	10	30
6	12	42
7	14	56
8	16	72
9	18	90

Double-Step Staircase tile pattern

One way to look into the totals for the Double-Step Staircase is to move tiles from taller columns to shorter ones to get the same number in all columns (see the diagram).

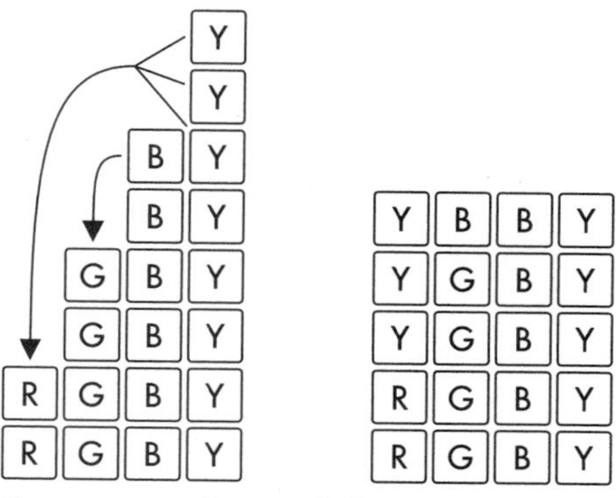

If you put 3 yellows with the reds and 1 blue with the greens, you make 4 columns of 5 tiles.

After step 2, we have an average of 3 tiles in each column and a total of 6 (2×3); after step 3 we have an average of 4 tiles in each column and a total of 12 (3×4); after step 4 we have an average of 5 tiles with a total of 20 (4×5). Each total continues to be the product of the step number and the step number plus 1.

If any students find a pattern for the Double-Step Staircase and present their method to the class, you might show this averaging as another way to approach the pattern; or you might show it just to the students who worked out their own approach.

Sessions 3 and 4

Four Growing Tile Patterns

Materials

- Student Sheet 2 (four pages: 1 set per student plus transparencies of the Twos Tower and Squares patterns)
- Square tiles in mixed colors (1 bucket per 6–8 students)
- Markers, pencils, or crayons in the tile colors
- Centimeter graph paper (2 sheets per group of 3–4, and 1 per student, homework)
- Overhead projector
- Transparent tiles for overhead (optional)
- Large paper for posters, scissors, tape

What Happens

Students analyze four tile patterns with these growing totals: 2, 4, 6, 8,...; 1, 4, 9, 16,...; 1, 3, 6, 10,...; and 2, 6, 14, 30,.... In tables and on graphs, students show how the tile patterns and the number series grow (whether at a steady or accelerating rate). Small groups investigate one of the four patterns in depth. They make a poster that shows the tile pattern shapes, a table, two graphs, and a written description of how their pattern grows. Student work focuses on:

- extending tile patterns
- making tables and graphs of step size and totals
- comparing graph shapes and explaining the differences
- investigating number sequences
- making several growing designs to fit the same table of numbers
- finding the *n*th term of a series
- Student Sheet 1 (1 per student, homework)

Ten-Minute Math: Nearest Answer In any spare 10 minutes outside of the regular math instruction period, do this Nearest Answer activity for ongoing practice with percents.

Select two or three Nearest Answer percent problems from p. 106 or prepare a few of your own, including answer choices, for display on a transparency or at the board. For example:

26% of 77 =	20	40	50	100
38% of 21 =	5	8	11	60

Keeping the answer choices covered at first, show the problem for 20 to 30 seconds. Students work mentally, using familiar percents to help them arrive at an estimated answer for the problem. For example, they might think:

26% of 77 is about 25% (1/4) of 80, or 20.

38% of 21 is about 40% (2/5) of 20, or a bit larger than 33 1/3% (1/3) of 21—close to 8.

Uncover the answer choices. Students choose the answer they think is the closest. Ask a few students to share their strategies for estimating an answer.

For complete instructions and variations on this activity, see p. 105.

Activity

Making Tables and Graphs for Tile Patterns

Hand out the four pages of Student Sheet 2, Growing Tile Patterns, to each student and distribute the tiles for groups to share. Students extend each tile pattern and fill out the tables and graphs as they did in the last session.

Students will have the rest of this session and part of the next one to work on the four pages. Caution them not to rush, and encourage them to consult with their classmates frequently or even work together throughout the activity. Students tend to be more careful with this work when they have to explain their thinking and get the agreement of a partner.

Before you draw a graph, check with your neighbors to be sure you agree about the numbers you have written in the table.

Circulate to help as students get started. Refer to the **Teacher Note,** Four Patterns (p. 36), for information about each pattern. Check to be sure students are extending the patterns correctly. For the Staircase pattern, some students may predict totals of 1, 3, 6, 9, 12,... instead of 1, 3, 6, 10, 15,... because from a quick look at the table, they have decided the totals are multiples of 3.

As students work, find out how they are filling in the tables. Pick out a row of a table and ask a student to tell you what the three entries mean. For example, in the fifth step of the Squares pattern, we use 9 new tiles (4 each for the top and side of the 4-by-4 square and 1 for the corner), and have a total of 25 tiles (5 rows of 5). In the fifth step of the Staircase, we take 5 (the same as the step number) and have a total of 15 (1 + 2 + 3 + 4 + 5).

When students seem ready, ask them if they can think of a way to get the *step size* and the *total* for any number without filling out all the steps on the table. Ask about steps beyond those on the table.

In the Squares pattern, how many new tiles will you use in the twentieth step? What will the total be? How did you figure it out?

The **Dialogue Box,** Finding General Rules for the Staircase Patterns (p. 38), illustrates how students in one class began thinking about general rules.

Although the names we have given to these four patterns suggest specific shapes, let students know that the tile patterns they are investigating do not *have* to look exactly like these. In the next session, they will have a chance to build different growing shapes that fit these same tables of numbers.

For example, the Staircase pattern is the "triangle" numbers (1, 3, 6, 10, 15,...), growing in larger and larger steps each time. When we make a column of each new step, the pattern looks like a staircase. Students often arrange the same numbers of tiles to make diagonals instead.

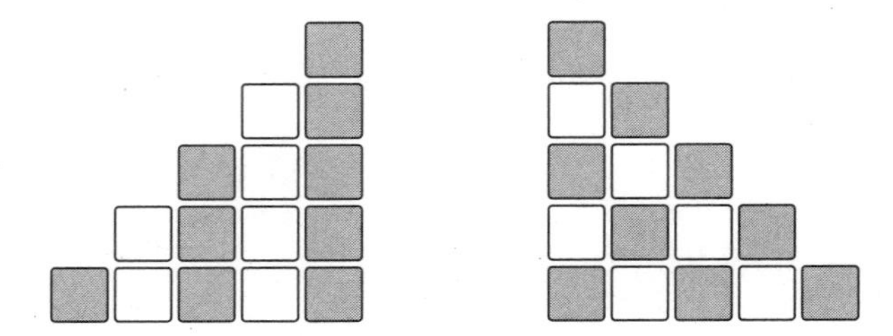

Rearranging the tiles in a pattern might help students find a general rule. However, when they do this, they must be sure to use the same step sizes in the same order as in the original design.

Activity

Comparing Graph Shapes

Toward the end of Session 3, bring students together to discuss their work on the four patterns on Student Sheet 2. Put the transparency of the Twos Tower on the overhead. With students' guidance, continue the pattern on the grid and fill out the table and graphs. Then ask about the shapes of the two graphs.

How do the different graphs show growth? Does either graph show no growth? Why is that? Which graph shows steady growth?

Do the same for the Squares pattern, and ask students to compare the two graphs for this pattern.

Which graph shows growth that is accelerating—growing faster and faster? How does that look different from the graph that shows steady growth?

This analysis is the heart of this activity. All these problems have a fixed solution, but how students think about the patterns and compare them shows their understanding of the important concepts presented here.

Suggest that students write under each graph how the pattern is changing: not growing at all, growing steadily, or growing faster and faster.

Which graphs look similar? Why are they similar?

Ask students to write about any general rules they have figured out. They can write on the backs of the student sheet pages if they need more room.

How would you figure out the step size and total for the next step? How would you figure out the step size and total for a much bigger step number, such as 20 or 100? Write anything you can say about this.

❖ **Tip for the Linguistically Diverse Classroom** Pair English-proficient students with those for whom English is a second language and let them do this task together.

Activity

Investigating a Pattern in Depth

Write on the board the first four numbers of the step-size sequence for the four patterns students have been exploring:

Twos Tower	Squares	Staircase	Doubling
2, 2, 2, 2,...	1, 3, 5, 7,...	1, 2, 3, 4,...	2, 4, 8, 16,...

Ask students how you could continue each series for a few more steps.

Explain that students will be working in groups to continue to investigate one of these four number series. You will tell them which series their group will explore. They will make a poster displaying information about the series.

One of the things each group must show on their poster is different tile patterns that grow with the same series of numbers. To illustrate this, recall the Double-Step Staircase from an earlier session, with the step-size number series 2, 4, 6, 8... Here are three other designs that grow in the same number pattern:

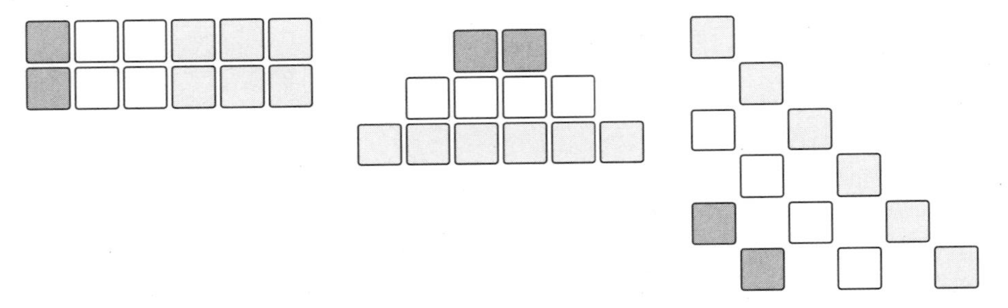

You might illustrate these alternative patterns at the overhead with transparent tiles or draw them on transparent grids.

Provide each group with large paper for the poster background. Also make available crayons, markers, or colored pencils in the tile colors, centimeter graph paper, scissors, and tape. Students can use these to make tile patterns and graphs, cutting them out and taping them on the poster. Tables might be drawn directly on the poster.

List on the board or on chart paper some of the things students could include on their posters to show how their series changes as it grows.

On your poster, try to teach something about growing tile patterns to someone who is not in this class. Think about how you might teach a parent or a brother or sister. Your poster might include the following:

- **Pictures of several tile patterns that grow with this same series (the one on the student sheet plus others).**
- **A table, showing number of new tiles added (step size) and total tiles used so far, out to 10 steps. If you know how to find step size and/or total for higher numbers, such as 20 or 50 or 100, show them on your table, too.**
- **A graph of step size vs. step number**
- **A graph of total (tiles) so far vs. step number**
- **A written description of how the pattern grows, using your tile designs, table, and graphs. Does it grow taller? wider? both? Does the pattern increase in size steadily, or faster and faster, or slower and slower? How can you tell by looking at the table? How can you tell by looking at the graphs?**
- **A description of any methods you used to find next numbers in the table, or to find step size or totals if you skipped to higher numbers.**

❖ **Tip for the Linguistically Diverse Classroom** Students might use symbols to represent steady growth, faster and faster growth, and slower and slower growth. They could work out their own symbols, making sure you understand, or work as a group to establish symbols they are satisfied with. Students can circle key parts of their graphs and tables to show what the symbols relate to. The whole class will give thought to this kind of symbolic representation in Investigation 2.

Assign the four patterns so that each pattern is investigated by one or more groups. You may want to take into account level of difficulty (from easiest to most difficult: Twos Tower, Squares, Doubling, and Staircase). If some students are reluctant to do the Twos Tower because it seems boring, or if you need *more* easy patterns, suggest students do a Ones Tower, a Twos Tower, and a Threes Tower, and compare the graphs for all three. If you need an additional difficult series, include the Double-Step Staircase from Session 1 or suggest that students investigate a Triple-Step staircase (adding 3, 6, 9, 12,...).

As groups begin work on their poster, check to be sure they understand the task and that they are all participating and making an effort to share the work. Once students have made a table up to the tenth step, suggest they make a row for the twentieth step and try to fill it in without counting all the steps between 10 and 20. This may be too difficult for some of the patterns, so avoid pressuring students to do this. If they can explain how to find the twentieth step, ask them how they could find the numbers for any step (the nth step). To demonstrate, they write in their own words how to get the step size and the total for the nth step.

Activity

Presenting the Posters

At the end of these sessions, or when all the groups have had time to make their posters, arrange for groups to present their posters to the whole class. Groups who investigated the same series might make their presentations together, taking turns showing what they found out about that series.

Display the posters in the hall where other classes can see them.

Sessions 3 and 4 Follow-Up

- After Session 3, students take home their work on Student Sheet 2 and a sheet of centimeter graph paper. They choose one pattern from the student sheet and draw different growing patterns that also fit the numbers in the table. Students may be able to use some of their new tile patterns on their posters in Session 4. They must be sure to bring all of Student Sheet 2 back to class for discussion during Session 4.
- After Session 4, students take home their folder of work from Investigation 1. They spend time checking it over, making corrections and finishing any incomplete pages. They might select their favorite tile pattern to keep in their work portfolio.

Tile Designs to Fit Descriptions Write on the board phrases that describe ways that a growing tile pattern can change, either in its step size or in its totals. For example:

Grows steadily   Grows faster and faster   Grows slower and slower

Shrinks steadily   Shrinks slower and slower   Shrinks faster and faster

Grows and then shrinks   Oscillates between growing and shrinking

Distribute centimeter graph paper. Students use this to make (1) a tile design whose total or step size changes in a way that fits one of the descriptions, and (2) a graph that fits the same description. They then write a description of the overall pattern they have made.

Teacher Note — Four Patterns

The Two Towers, Squares, Staircase, and Doubling patterns are intriguing examples of the fusion between geometry and numbers. As students experiment with them, they will see the patterns from many different perspectives. For some patterns, they will be able to verbalize a general rule for finding step size and total at the nth step; for other patterns, this will be too difficult.

Twos Tower The general rules for the Twos Tower are similar to the rules for the Threes Tower presented in Session 1. The pattern has a horizontal line graph for the step size (step size = 2 always) and a straight line graph for the totals (total = 2 × step number).

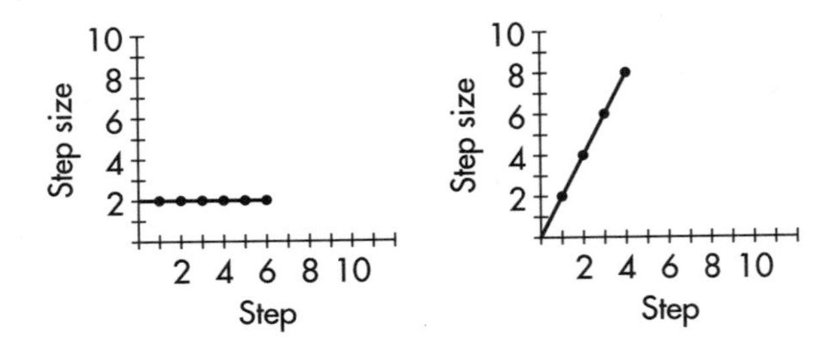

Step number	New tiles (Step size)	Total so far
1	2	2
2	2	4
3	2	6
4	2	8
step number n	always 2 2	twice the step number $2n$

Students are likely to figure out the general rules for this pattern. You can use the totals in this example to help students understand the idea of finding a general rule that can be used for any step.

If you know the step number, what calculation can you do to find the total? What is the total for step 100? for step 50? How did you figure that out?

Squares In the Squares pattern, the step size (number of tiles added each time) grows at a steady rate, while the total number of tiles grows at an increasing rate.

Step number	New tiles (Step size)	Total so far
1	1	1
2	3	4
3	5	9
4	7	16
step number n	1 less than twice the step number $2n - 1$	step number times itself n^2

Finding the totals (the step number multiplied times itself, or squared) may be easier than finding the step size. Refer the students to the actual tiles, or to the picture on grid paper, to explore the number of tiles added at each step.

When you add new tiles to make a bigger square, where do you put them? How can you know how many you need by looking at the square that is already there?

Staircase For the Staircase pattern, the numbers for the totals (1, 3, 6, 10, 15,...) are often called triangle (or triangular) numbers because the steps of the pattern can be arranged as a triangle.

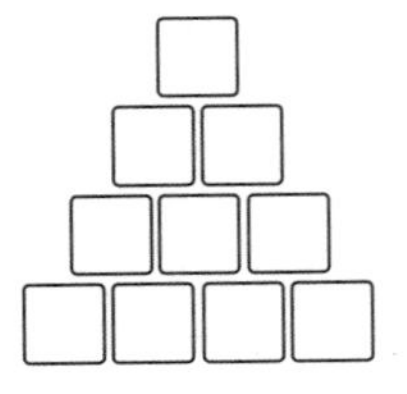

Continued on next page

Step number	New tiles (Step size)	Total so far
1	1	1
2	2	3
3	3	6
4	4	10
step number n	same as step number n	multiply the step number by the *next* number, then divide by 2 $\frac{n \times (n + 1)}{2}$

As is true of the Squares pattern, the total number in the Staircase pattern grows at an increasing rate, while the step size grows steadily. The number of tiles added at any step is the same as the step number, but the general rule for the total is much more difficult to figure out, and students should not be expected to find it.

If some students want to figure out the general rule for the Staircase total, suggest that they first try finding the general rule for the total in the Double-Step Staircase, then compare the numbers for the two patterns. The numbers for step size and total in this single-step Staircase are half as big. In the Double-Step Staircase, the total was the step number multiplied by the *next* step number: $n \times (n + 1)$.

Alternatively, suggest that students investigate the size of the rectangle that the Staircase is half of. (It is half of a rectangle made by *that* step number multiplied by the next step number; for example, if you put two of the staircases with three steps together, you can make a 4 × 3 rectangle.)

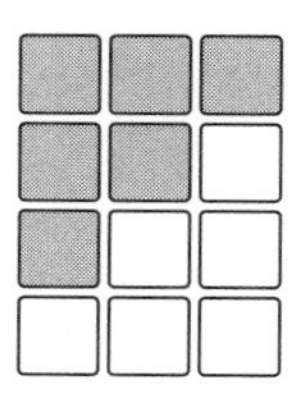

Doubling The Doubling pattern is an example of one for which both the total and the step size grow exponentially. A related example of exponential growth would be as follows: One person tells a story to two others, and each of them tells two others, and each of them tells two others, and so on. The step-size numbers are powers of two: $2 = 2^1$, $2 \times 2 = 2^2$, $2 \times 2 \times 2 = 2^3$, ..., 2^n.

Step number	New tiles (Step size)	Total so far
1	2	2
2	4	6
3	8	14
4	16	30
step number n	2 raised to the power of the step number 2^n	2 raised to the power of the *next* step number, then subtract 2 $2^{n+1} - 2$

Both general rules for the Doubling table are too difficult for most fifth grade students to figure out on their own. However, once students see the pattern of multiplying by 2 again and again, they may enjoy it and want to go on doubling to get higher and higher step sizes.

You might take suggestions from students for a title for this shape. The name "Doubling" describes what is happening to the step size, but is not descriptive of the overall shape.

DIALOGUE BOX

Finding General Rules for the Staircase Patterns

These students have been working on Student Sheet 2, Growing Tile Patterns. The teacher stops them to hold a brief discussion on the Staircase pattern. When asked how they are deciding what numbers to fill in, the students offer a variety of approaches and generalizations.

Staircase

Step number	New tiles (Step size)	Total so far
1	1	1
2	2	3
3	3	6
4	4	10
5	5	15
6	6	21

Tai: The step size is like the same as the step number, because you take one more every time. It starts from 1, and then 2, and 3, and then 4, and then 5, and then 6, and then 7, and 8, and then 9. Whoa [*indicating there's no more room on the table*]—way overboard!

Jeff: The total is 1, 3, 6, 9, and stuff. Threes, remember?

Look at your tile pattern again. How many tiles are there, total, after you add yellow *[the fourth step]*? How many after you add red again? Count carefully to check your totals. Amy Lynn?

Amy Lynn: First I saw a pattern. The total just goes up by 1, 2, 3, 4, 5, 6, 7. See the 1; it has no number on top of it, so it just equals 1. And then 1 from here *[indicates total column]* with 2 from here *[indicates step size column]* equals 3, and then 3 and 3 equals 6, so you just keep adding [the previous total to the new step size].

Katrina: Another pattern we saw is, look up here [*indicates totals*]. To get from 1 to 3 is 2. Then it'd take 3 to get to 6, and then 4 to get 10. So add 2 to get from 1 to 3, then add 3 to get from 3 to 6, then add 4, and then add 5. And it keeps going.

Anything else about the totals?

Noah: It'll be curved on the graph. 1, 3, 6, 10, 15 and then it would just go bigger and bigger off the top.

Is there a way to find the numbers at the tenth step?

Marcus: I wrote n plus previous n's equal the total.

If you know this pattern, you can do that, absolutely. I guess my question is, What if it were the 126th step? Is there a way to figure it out no matter what step it is?

Yu-Wei: Yeah. You'd hafta know the step before. I sorta thought I found a pattern too, but it doesn't really work because you need to know what the last step was.

Articulating a general rule is difficult for these students. Following the rule that they generated would require a lot of computation for higher-numbered steps. The teacher thinks some of the students are ready to investigate a complicated general rule, but not one as difficult as the Staircase pattern.

While the rest of the class continues to work on the patterns on Student Sheet 2, the teacher has a special project for those who are ready: Look for a general rule for the Double-Step Staircase pattern they talked about in the previous session. After about half an hour, one pair finds a general rule that doesn't quite work.

Continued on next page

Double-Step Staircase

Step number	New tiles (Step size)	Total so far
1	2	2
2	4	6
3	6	12
4	8	20
5	10	30
6	12	42
7	14	56

Jasmine: We think we've got it. Yeah, because *[without looking at the table]* 1 times 2 is 2. Then 2 times 3 is 6 for the total up to that row. 3 times 4 is 12. 4 times 5 is 20. 5 times 6 is 30. 6 times 7 is 42. 7 times 8 is 56. And 8 times whatever. 9 times 10 would be 90.

What about the twentieth step? What arithmetic should I do to find the total after the twentieth step?

Marcus: 19 times 20'd be...

[The teacher covers the seventh row with her hand.] **What if *n* is 7?**

Marcus: 42.

Jasmine *[looking at the now uncovered table]:* No, it's 56. Oh, it's 7 times 8.

Take some more time to think about the rule.

Jasmine *[a little later]:* We got it this time. Say the step number is 50, you do 50 times 51, which'd be... 50 times 50 is 2500 and then 1 more 50'd be 2550. So if 50 is n, then 51 is n + 1, so it's n times n plus 1.

How did you figure that out?

Jasmine: Well, before we were doing the step number times the number on the line above it, but we were going the wrong way. You have to do it by the *next* number. That's the one below it on the table, but it comes after it. It's a higher number.

INVESTIGATION 2

Motion Stories, Graphs, and Tables

What Happens

Session 1: Describing Changing Speeds Students plan and act out trips of varying speeds along a straight line track. They develop ways to record the trips, without using words, clearly enough so that someone who has not seen the action can describe the trip. Students interpret and critique one another's representations.

Session 2: From Beanbags to Tables Students record their trips along a track by dropping beanbags at two-second intervals. They make tables and diagrams showing where the beanbags landed. Students then exchange their work and try to describe each other's trips by interpreting the tables and diagrams.

Session 3: Tables for Stories Each student makes a table for one of three motion stories, showing where beanbags might have dropped at regular time intervals. Students exchange tables and decide which motion story belongs with the table they receive.

Session 4: Graphs for Tables Working from the table they made in Session 3, students make a line graph of distance versus time. As they did with the tables, they exchange graphs and decide which motion story best fits the graph they receive. Students compare graph shapes that represent the same story to identify common characteristics.

Session 5: Stories, Tables, and Graphs Students match motion stories, tables, and graph shapes that describe the same trip. They make a table to go with one of the graphs and write a motion story to go with another.

Mathematical Emphasis

- Exploring relationships among distance, time, and speed
- Exploring irregular increase and decrease in step size and total
- Exploring ways that speed, time, and distance can be represented with tables, graphs, stories, and informal representations
- Interpreting intervals in a table as reflecting speed
- Interpreting steepness in a distance vs. time graph as reflecting speed
- Associating tables, graphs, and stories of the same event

What to Plan Ahead of Time

Materials

- Wide (2-inch) masking tape: about 50 meters (Session 1)
- Unlined paper: 4–5 sheets per student (Sessions 1, 3, and 5)
- Markers, crayons, or colored pencils (Sessions 1, 3, and 5)
- Timepieces that show seconds: 1 per 4–6 students (Sessions 1–2)
- Beanbags 1–2 inches in diameter, or other small objects that stay in place when dropped (see Other Preparation). Place about 12 of one color in a shallow container for each 4–6 students. (Session 2)
- Stick-on notes: 1 pad (Sessions 3–4)
- Overhead projector (Sessions 3–4)
- Scissors and tape or glue sticks: 1 per pair (Session 5)

Other Preparation

- Before Session 1, find four or more places where you can lay out straight lines (or tracks) of masking tape, 8 to 10 meters long. One track must be in the classroom; others can be in the hall or outside. Students will work at the lines in groups of 4–6. Mark (or have students mark) these tracks by meters and half-meters. Number the whole meters clearly.

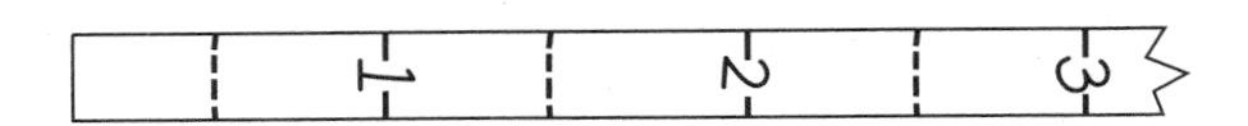

- Before Session 2, prepare small "beanbags" in two different colors. You can wrap beans in scraps of fabric, closed with staples or stitching. Or fill small plastic bags, doubled for strength, with beans or sand; close them with rubber bands.
- Plan some wall space to display students' tables and graphs in Session 3. Write the three motion stories from Student Sheet 5, in large lettering, as titles for the displays.
- Duplicate student sheets and teaching resources (located at the end of this unit) as follows:

For Session 2

Student Sheet 3, Template for Tables (p. 126): 1 per pair

Student Sheet 4, Height of a Girl (p. 127): 1 per student (homework), plus 1 transparency (for Session 3)

For Session 3

Student Sheet 5, Three Motion Stories (p. 128): 1 per student

Student Sheet 6, Graph of a Trip (p. 129): 1 per student (homework), plus 1 transparency (for Session 4)

For Session 4

Student Sheet 7, Graph Template (p. 130): 1 per student, plus some extras

Centimeter graph paper (p. 151): 1 per student (homework)

For Session 5

Student Sheet 8, Matching Stories, Tables, and Graphs (pp. 131–132): 1 per student

Session 1

Describing Changing Speeds

Materials

- Masking tape tracks laid out on floor (1 per 4–6 students)
- Timepiece that shows seconds (1 per 4–6 students)
- Unlined paper; markers, crayons, or colored pencils

What Happens

Students plan and act out trips of varying speeds along a straight line track. They develop ways to record the trips, without using words, clearly enough so that someone who has not seen the action can describe the trip. Students interpret and critique one another's representations. Student work focuses on:

- describing trips with changing speed along a straight line path
- making visual representations to describe trips
- interpreting representations of trips

Ten-Minute Math: Nearest Answer Continue to do the Ten-Minute Math activity, Nearest Answer, over the next few days. Change the focus from percents to fractions, mixed numbers, and decimals. Remember, Ten-Minute Math activities are done in any spare 10 minutes outside of math time, perhaps just before or after lunch, recess, or gym.

Follow the same procedure as for Nearest Answer percent problems, but choose from the fraction, mixed number, and decimal problems suggested on p. 106 or make up your own. Students are to round the numbers in the problems to the nearest whole or "landmark" numbers and estimate.

For complete instructions and variations on this activity, see p. 106.

Activity

Describing Fast and Slow Trips

Students sit where they will be able to see the motion of a person traveling along a "track" (one of the lines of masking tape you have laid out and marked by half and whole meters). Choose a volunteer to demonstrate the activity. Explain that students will move alongside the tape according to a plan that involves changing their speed.

Prepare a plan for the first trip and tell it to *only* the student who will act it out. For example: "Walk slowly about halfway, then stop for a few seconds, then run to the end."

While Sofia moves along this track, watch carefully so that afterward you can describe the trip to me. In particular, pay attention to *how* her speed changes, and *where* it changes. Watch the meter marks on the track. Where does she go fast? Where does she go slow? Where does she change speeds?

After your volunteer acts out the plan, students take turns describing the speeds in the trip.

Tell the story of the different speeds that Sofia went. Be as exact as you can. Tell at which meters she changed speeds.

If there are different interpretations, the same student acts out the trip again, and the rest of the students decide which of their descriptions is most accurate. Students discuss the trip until they agree.

For more practice, choose another volunteer to do a different trip. For example: "Run for three steps, then stop for 5 seconds, run three more steps, stop for 5 seconds, then walk slowly to the end." Again, students share their descriptions of the trip.

Activity

Representing Changes of Speed

Working in pairs or groups of three, students plan a trip along the tape track, different from the trips you just demonstrated. Instead of a verbal description, they create a visual representation of their trip. Explain the activity to the whole class, writing key instructions on the board.

With your partner, secretly plan a trip along the track. Be sure your trip has some changes of speed. Then invent a way of showing the changing speeds on paper, *without words* and *without a key*. You may make a table if you wish.

When you have shown your trip on paper, exchange papers with another pair. Try to figure out each other's trips.

While students are planning their trips, the pair that they will exchange with should not be watching. Thus, if two or three pairs must share a track, they should find a pair working at another track to exchange representations with. If two pairs working at the same tape will be exchanging representations, they must be careful to keep their trips secret.

Allow no more than 10 minutes for making the representations. Suggest that students keep them simple, avoiding elaborate drawings or symbols that take a long time to make. While students are working, observe and try to understand the trip they are describing. Ask for more clarity when you think it is needed, but do not expect or encourage any kind of conventional graphs.

Although students made conventional graphs to show the growth of tile patterns in Investigation 1, they are not expected to make similar graphs for this new situation. See the **Teacher Note,** Invented Representations of Trips (p. 46), for some observations about the issues students focus on and the kind of representations they typically make.

In other sessions in this investigation, students will use line graphs and tables to represent trips, just as they did for the growing tile patterns. But in this activity, students should be thinking about their *own* ways to communicate distance and speed. This is not an easy task, and it will help students appreciate the efficiency of the standard line graphs they will be making later, although they may regret the loss of some of the detail they included in their own more personalized graphs.

Tell students when time is almost up so they can finish and be ready to show their representation to the other pair.

Describing the Other Pair's Trip After each pair has had some time to discuss and interpret the representation they have received, they take turns trying to describe *and act out* the motion of the trip along the track. The pair who produced the drawing must refrain from giving any verbal hints, but they can write labels, make tables, or draw directly on their representations to make them clearer for the readers.

When both pairs have done their best to interpret each other's trip, students fix their representations to make them clearer.

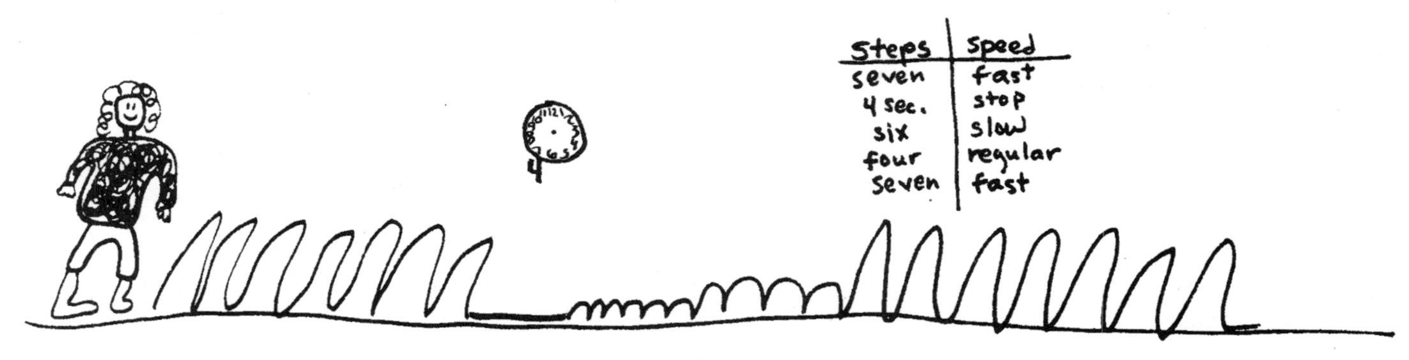

Large-Group Sharing If there is time, arrange for students to come together in groups of eight to ten students (or as a whole class, if the class is small) to look at one another's trip representations. Student pairs take turns presenting their diagrams or drawings. They ask other students to guess what the representations mean and then explain what they were trying to communicate. Suggest that students look for good ways to show length of time, distance, changes of speed, and stopping. Look especially for ways of showing speed that can be varied easily to show different speeds.

Keep the tape tracks in place for the next session.

Displaying Representations as a Puzzle You might make a display of the students' trip representations in the hall, with all of the corresponding "stories" typed up on a separate piece of paper. Students from other classes can guess which story goes with which representation. They could write their guesses on a piece of paper tacked to the wall for this purpose.

Session 1 Follow-Up

Students make a drawing of a straight line trip with changing speeds and ask someone at home to act it out. They make changes, if necessary, to make the trip easier to interpret, and ask the person to try again until the reader can interpret the trip correctly, working from the representation alone.

Students write a few notes on their representation about what the person did and didn't understand, and describe the changes they made to their representation.

❖ **Tip for the Linguistically Diverse Classroom** Students who are not comfortable writing in English might write their notes in their primary language.

Teacher Note — Invented Representations of Trips

Fifth grade students have many intuitions about and experiences with motion. When they are asked to represent motion on paper, they express what they regard as the important aspects of a particular trip. Students tend to include and combine the following elements in their work:

- starting position
- speed (fast, slow, steady)
- reference points (halfway, or at the sink)
- length of time
- stops along the way
- direction of the trip (forward, backward)
- final position

Students often combine pictorial and graphic aspects of motion in their representations. Some typical characteristics of students' spontaneous work include the following:

- They often represent a trip along a line drawn as if the line were the path (and may include a stick figure of a person taking the trip). Even students who have worked with two-dimensional coordinate graphs in other contexts do not think to use that technique here. They tend to use a line as one dimension to show distance, and add symbols of various sorts to show speeds. Most use a special symbol, such as a stop sign, to show no motion.
- They sometimes use discrete symbols or colors to show different speeds. Students may develop a scheme that can vary for different speeds: for example, a wavy line with higher or lower waves, larger and smaller dots, or footprints farther apart or closer together to show faster or slower motion.
- Students are most likely to use a variety of symbols, such as colors, that do not by themselves indicate speeds, and then provide a key to show meanings. They seldom use a scale to indicate speeds; if they use numbers, it is usually to indicate miles per hour.

Whether they use graphics, pictures, or both, students are symbolizing aspects of motion that they consider important in describing and understanding their trip.

One important goal of this activity is that students consider representations as a form of communication. Thinking of their graphing and use of symbols as a way to inform someone else about their trip is likely to be new for many students. By grappling with the task of representing motion, students will encounter for themselves the many problems that people have attempted to deal with by using conventional graphs and number tables.

Like language, representations exist to inform and communicate; and, like language, they can be ambiguous. In fact, when we step back to look at our representations, we may find some surprises. There may be aspects of the trip we thought we'd expressed that are not really discernible. For example, drawing a line quickly does not leave an indication of fast motion. Through revising their representations, students make them more consistent and often improve their ability to communicate. For example, in the work shown below, the students at first used numbers only to show the order of events. Later they decided to put numbers on the stop sign symbols to show how long the stops were between each set of 3 steps (the stops increase by one second each time).

From Beanbags to Tables

What Happens

Students record their trips along a track by dropping beanbags at two-second intervals. They make tables and diagrams showing where the beanbags landed. Students then exchange their work and try to describe each other's trips by interpreting the tables and diagrams. Their work focuses on:

- interpreting spacing of dropped objects as speed
- collecting and recording data in regular time intervals
- making a table of time and distance
- interpreting a table that shows accumulated distance and time
- measuring or computing intervals

Materials

- Timepieces that show seconds (1 per 4–6 students)
- Student Sheet 3 (1 per pair)
- Student Sheet 4 (1 per student, homework)
- Beanbags in two colors (12 of one color per 4–6 students, placed in shallow containers)

Activity

Comparing Fast and Slow Trips

Ask students to gather around a marked tape track. Invite two volunteers to demonstrate the activity. Each volunteer takes a container of beanbags (each a different color so you can identify which beanbags have been dropped by each student).

Explain to the two volunteers, out of other students' hearing, that when you say "start," they will both move along the track, starting at the same end, one on each side of the track, but at different speeds. Together they agree who will walk slowly and who will walk quickly. Every time you say "drop," they will both drop a beanbag from their container. Both stop when they reach the end of the tape.

Before you begin the trips, ask the other students to turn their backs or close their eyes, but to listen carefully when you say "start" and try to figure out what you and the two demonstrators are doing. You will need a timepiece that shows seconds as you run the trips.

After you say "start," say "drop" every 2 seconds. Continue this until both students have reached the end of the tape.

Because there will be a delay in the students' reaction time, the interval from the start to the first beanbag drop will be longer than the other intervals, which should be quite regular. The fast walker's intervals (space between beanbags) will be longer than the slow walker's intervals.

At the end of the trips, the rest of the class can open their eyes and look at the track.

What do you think we were doing?

After some students guess, tell them that the demonstrators dropped a beanbag every 2 seconds.

How do you think Corey and Heather were moving along the tape? Which line of beanbags was dropped by someone moving fast? Which was dropped by someone moving slowly? How do you know?

Some students may find it confusing that the person moving fast dropped fewer beanbags with larger spaces in between (•   •   •   •   •) than the person moving slowly (• • • • • • • • •). One way to make sense of this is to note that the number of beanbags is a measure of time. The greater number for the slower trip shows that it took *more* time for the slower person to go the whole length; that person's time was *more* multiples of 2 seconds.

How can we figure out how long the faster walker took to go the whole length? How many seconds altogether? (It may not be possible to answer precisely; for example, in the table below, the fast walker took somewhere between 6 and 8 seconds to complete a 10-meter trip.)

How can we tell how long the slower walker took?

On the board, draw two tables like the ones below. As you fill in the 2-second intervals, students tell you the measures where beanbags were dropped, to the nearest half meter. Record these distances in order, first for the slow trip and then for the fast trip. You might have the demonstrators re-enact each trip in 2-second segments as you record.

After the first 2 seconds, where are you? *[The walker moves to the first beanbag at the same speed as in the original trip and, with help from other students, tells you what distances to record.]* **In another 2 seconds, where are you?** *[The walker moves to the next beanbag, other students decide what the measure is, and you record.]*

Continue to the end of the trip. The measurements may resemble these, although they are unlikely to be this regular:

Slower Trip	
Time	Total distance so far
2 sec.	1.5 m
4 sec.	2.5 m
6 sec.	3.5 m
8 sec.	4.5 m
10 sec.	5.5 m
etc.	etc.

Faster Trip	
Time	Total distance so far
2 sec.	3 m
4 sec.	6 m
6 sec.	9 m
8 sec.	end

Corey, the slower walker, dropped more beanbags because he went slower. How far apart are most of his beanbags? About how far did he usually travel in 2 seconds?

As students figure the length of the intervals (space) between beanbags for the slower walker, write them in just to the right of the table (as shown on the next page).

Heather, the faster walker, dropped fewer beanbags. About how far apart are her beanbags?

Repeat the process of figuring and noting the intervals between beanbags, this time for the faster walker.

Slower Trip		
Time	Total distance so far	Space between beanbags
2 sec.	1.5 m	1.5m
4 sec.	2.5 m	1m
6 sec.	3.5 m	1m
8 sec.	4.5 m	1m
10 sec.	5.5 m	1m
etc.	etc.	etc.

Faster Trip		
Time	Total distance so far	Space between beanbags
2 sec.	3 m	3m
4 sec.	6 m	3m
6 sec.	9 m	3m
8 sec.	end	end

As needed, clarify what you are doing here by relating it to a car trip students might have taken. Ask them to imagine that they are taking a trip to a city that is about 4 hours' drive away. At the start of the trip, they reset the trip odometer on the car's dashboard to 0. After every hour, they look at it. At the end of the first hour, the odometer reads 40 miles; then at the end of the next hour, it reads 98 miles; after the third hour, 151 miles; and after the fourth hour, 175 miles.

Write the numbers in a table. Ask students to tell the story of when the car went fast and when it went slowly by finding the number of miles driven each hour. Ask students to speculate about why the speed on the trip might have varied this way. For example, perhaps in the first hour they drove along streets with lots of stoplights, so they went more slowly; then they got on the highway where they could travel faster for two hours; then in the last hour, maybe they stopped for a snack.

Car Trip		
Time	Total distance so far	miles each hour
1 hr.	40 mi.	40 mi.
2 hr.	98 mi.	58 mi.
3 hr.	151 mi.	53 mi.
4 hr.	175 mi.	24 mi.

Comparing Another Two Trips Invite two students who are unsure of how to interpret the close-together and far-apart beanbags to make another trip along the tape track. Again, secretly tell the walkers how they are to move: One will walk quickly but then stop for 4 to 6 seconds and drop two or three beanbags on the same spot; the other will change speed in the middle of the trip (say, start slowly and end quickly).

For these trips, either you or a student can time and say "drop" every 2 seconds. The rest of the students again look away or close their eyes while the trips are carried out.

After the trips, ask students to read the measures where the beanbags landed. Demonstrate again how to fill out tables with the intervals of time (every 2 seconds) and the position reached, or total distance traveled so far (in meters), and how to compute the distance traveled in each 2-second interval by finding the difference between adjacent total distances.

Briefly make clear some conventions about tables:

- **When you make tables, you start the story at the top and read down.**
- **Tables can be two or more columns. Each column needs a title telling what the numbers in it mean.**
- **Across any row of a table, the numbers are related. For example, the table for the slower walker shows that after 6 seconds, he had reached the ___-meter mark. Where was the faster walker at 6 seconds?**

See the **Teacher Note,** Tables That Show Changing Speeds (p. 54), for a discussion of the tables students typically make on their own compared to the kind of table you are teaching them to make.

Activity

Marking and Guessing Trips

For this activity, students practice making trips and guessing speeds from the placement of dropped beanbags. They work in small groups at the different tape tracks in and outside of the classroom. Designate a Group A and a Group B at each track, with two or three students in each group.

At the tape tracks, the students will first practice timing, dropping beanbags, and guessing speeds from the placement of the beanbags; then they will make tables from which others can guess the relative speeds in their trip. Four to six students (two pairs or groups of three) will work together at one tape, taking turns acting out trips and guessing speeds. Then each pair or three will make a table of a trip for other students to interpret.

The two groups at each track will take turns. When it's your turn, your group makes up a trip, changing speeds along the way and dropping beanbags every 2 seconds. One person in each group will be the timer; you will say "drop" every 2 seconds. Only one traveler will take the trip and drop the beanbags. Everyone in the other group will look away while you do your trip.

When the trip is finished, the other group looks at where your dropped beanbags are and guesses where you moved fast, where you moved slowly, and whether you stopped anywhere. Take turns doing this until you are good at guessing how a trip went.

Take some time to plan your trips, secretly, before you act them out. Each group makes at least one trip and guesses at least once. Practice together reading the meters where the beanbags landed.

Be sure every group has a watch that shows seconds or a wall clock they can see clearly. When students start working, help anyone who is not clear on how to do the timing. Remind them as necessary that these trips are for just one person moving along a track, not two people as in the demonstration activity.

Observe the students to see what sense they are making of this activity.

- Can students read the measures where the beanbags are dropped?
- Can they determine the lengths of the intervals? (Many students have difficulty finding the size of an interval that is not whole meters, for example, from 3.5 meters to 5 meters.)
- Can students associate the length of the interval between beanbags with the speed at which the dropper was moving?
- Can they make sense of two beanbags dropped almost in the same place, showing that the dropper stopped or went very slowly?

Students will need to be able to read distances along the tape track and interpret the intervals between beanbags in order to make and interpret tables. To encourage students to relate the intervals between beanbags to varying speeds, ask:

About how long is the interval between beanbags when you walk slowly? Show me how you know.

About how long is the interval between beanbags when you run?

How do the beanbags look when they were dropped while someone was stopped?

If some students are not getting clear patterns, advise them to work on simpler motion stories; for example, "Go fast most of the way and then finish very slowly."

Encourage groups to move on to making tables (the next activity) as soon as they are ready.

Activity

Making Tables to Show Trips

As students are ready, give each group a copy of Student Sheet 3, Template for Tables. Each group plans a trip and makes a table for the trip that shows where the beanbags would land. They might make a table from the beanbags they dropped during a practice trip in the previous activity. Or, they can run another actual trip and record their measurements in the table. They could also invent a trip by laying out beanbags and recording where they place them along the tape track.

At least 10 minutes before the end of the session, gather everyone back in the classroom.

Now, each of you will exchange your table with another group. Before you give your table to someone else, be sure that you could figure out your own trip from your table. You might want to change slightly the places where the beanbags fell, to make the intervals more regular, if you think that would help someone understand your trip.

When each group has a data table they are happy with, they exchange it with another group. They guess how each other's trip goes and perhaps act it out. Then they discuss whether the trips were described or acted out correctly, and what they had intended their tables to convey.

The group who is trying to guess the trip may want to use the track on Student Sheet 3, marking with dots where the beanbags landed, to help them understand some details of the trip. If there is time, groups can do more than one trip.

Note: You will not need the tape tracks again in this investigation, but try to leave the one in the classroom in place for use at the beginning of Investigation 3. If you cannot leave it, plan to replace it for use in the first session of that investigation.

Session 2 Follow-Up

Students take home Student Sheet 4, Height of a Girl. They fill in the third column in the table and make a graph of the data. They use their work to write the story of how fast the girl grew at different ages.

❖ **Tip for the Linguistically Diverse Classroom** Instead of writing a story, students can create a pictorial or symbolic representation of the girl's growth.

Teacher Note — Tables That Show Changing Speeds

Showing Trips in Episodes (What Fifth Grade Students Typically Do) While we were developing this unit, we asked students to make tables for trips along the tape tracks without demonstrating a particular kind of table to them. These students had not worked with growing tile patterns (Investigation 1), so they hadn't seen those tables, either.

Most students spontaneously made tables like A, B, or C below (each shows the same trip). They thought of a trip in pieces as they planned it: "Shakita went real slow to 2 meters. Then she went fast for 4 meters and stopped. Then she went slow to the end." They timed and recorded each piece of the trip separately.

For tables A–C, they timed someone walking from 0 to 2 meters, then going faster from 2 meters to 6 meters, then stopping, then going slowly from 6 meters to 11 meters.

Table A

Meters	Time
2	4 sec.
4	2 sec.
0	3 sec.
5	10 sec.

Table B

Meters	Steps in a meter
2	4
4	1
0	0
5	4

Table C

Meters	Speed
2	slow
4	fast
0	stop
5	slow

It is important to avoid saying that any of these tables are "wrong." There are many ways of showing a trip with number tables, and there is no general criterion to say that one is right and another is wrong; it depends on what information we want to convey. The convention that we use in this unit (explained below) is convenient and commonly used, but it is one possibility among many.

Showing Accumulated Time and Distance (What We Are Teaching) Very few fifth graders spontaneously make a table based on one unit at regular intervals, such as the 2-second intervals in table D, or the 1-meter intervals in table E. We use a model with regular intervals so that students can interpret tables by finding time or distance in each interval, and so they can graph the numbers from the tables to show change over time.

In tables D and E, both the time and the distances accumulate so that reading across at any level shows the time passed and the distance traveled since the start. In this investigation, students make graphs that show changes in distance vs. time. In the third investigation, they also make graphs that show speed (expressed in terms of step size) vs. time.

By finding how much distance was traveled in a regular interval of time (2 seconds in table D) or how much time it took to go a certain distance (1 meter in table E), students can decide whether a person was moving slowly or quickly or was stopped.

Table D

Time	Total distance so far	Space between beanbags
2 sec.	1 m	1 m (slow)
4 sec.	2 m	1 m (slow)
6 sec.	6 m	4 m (fast)
8 sec.	6 m	0 m (stopped)
10 sec.	7 m	1 m (slow)

Table E

Distance	Total time so far	Time for last meter
1 m	2 sec.	2 sec. (slow)
2 m	4 sec.	2 sec. (slow)
3 m	4.5 sec.	0.5 sec. (fast)
4 m	5 sec.	0.5 sec. (fast)
5 m	5.5 sec.	0.5 sec. (fast)

Tables for Stories

What Happens

Each student makes a table for one of three motion stories, showing where beanbags might have dropped at regular time intervals. Students exchange tables and decide which motion story belongs with the table they receive. Student work focuses on:

- making tables of total distance in periods of 2 seconds
- analyzing intervals in tables to match tables to stories
- comparing tables of distances that fit the same story

Materials

- Transparency of Student Sheet 4
- Overhead projector
- Student Sheet 5 (1 per student)
- Three motion stories from Student Sheet 5, written large for wall display
- Student Sheet 6 (1 per student, homework)
- Stick-on notes (available)
- Unlined paper (1 sheet per student)

Activity

Interpreting the Table of Heights

For homework review, students spend a few minutes conferring with their neighbors, comparing their interpretations of the table for Height of a Girl (Student Sheet 4). Show the transparency of the student sheet while you discuss it with the class. You might uncover only a few lines of the table, say, ages 7, 8, 9, and 10, and ask students how they would describe the girl's growth during that period (steady growth, 2 inches each year). From these entries, what might they predict for the next few years if they hadn't already looked at the whole table?

Fill in the third column with the number of inches the girl grew each previous year (the first row is blank because we do not know her 6-year-old height, then 2, 2, 2,...). Slowly uncover more lines and ask students what change they see in her growth pattern. When you have uncovered the whole table, ask if the pattern makes sense.

Is this the way an actual person might grow? What is happening at the end?

Graph the heights from the table on the transparency. Use a ruler to connect the first five heights (48–56 inches) with a straight line; then shift the ruler to connect the next three heights (59–65 inches). Connect the last three points one by one; the result will be a "curve" showing that the girl's growth is slowing down.

When is the girl tallest? At what age does she reach her greatest height? How can you tell? How can you tell on the graph at what ages the girl grew fastest? Is this the same as when she was tallest?

How is the part of the graph that shows where she grew fastest different from the part that shows where she grew 2 inches a year? How can you tell that her growth slows down?

If no students point out the changes in steepness, point them out yourself.

When is the graph steepest? What does this mean?

Students should notice that when the girl is growing 3 inches a year, the graph is steeper—climbing faster—than when she is growing 2 inches a year. When she has almost reached her full height, her rate of growth slows down, and the graph becomes less steep. Don't worry if students don't seem to understand all the details at this time. Consider this an introduction to distance (or, in this case, *height*) vs. time graphs, in which changes of slope reflect changes in speed.

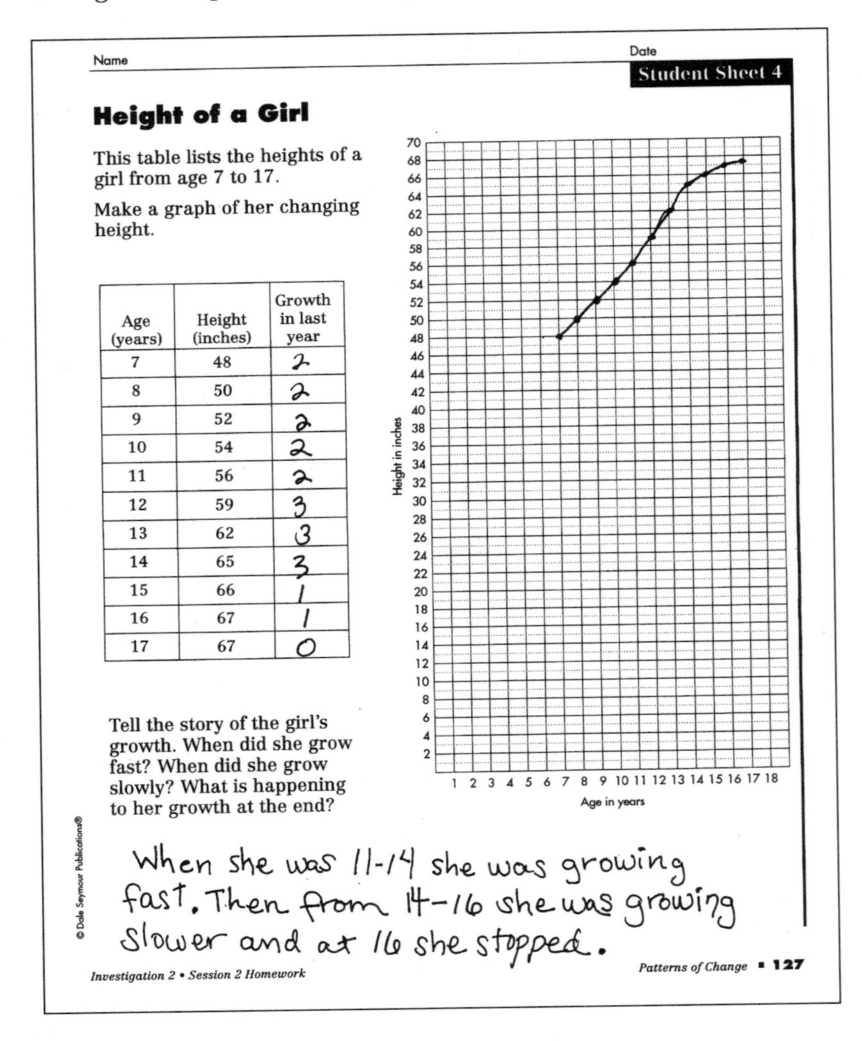

Name Date

Student Sheet 4

Height of a Girl

This table lists the heights of a girl from age 7 to 17.

Make a graph of her changing height.

Age (years)	Height (inches)	Growth in last year
7	48	2
8	50	2
9	52	2
10	54	2
11	56	2
12	59	3
13	62	3
14	65	3
15	66	1
16	67	1
17	67	0

Tell the story of the girl's growth. When did she grow fast? When did she grow slowly? What is happening to her growth at the end?

When she was 11-14 she was growing fast. Then from 14-16 she was growing slower and at 16 she stopped.

Investigation 2 • Session 2 Homework

Patterns of Change ▪ **127**

Activity

Teacher Checkpoint

Making Tables for Stories

Post the large prepared copies of the three motion stories. Leave space under each story for students to display the tables (and later graphs) they make to fit the stories. Give a copy of Student Sheet 5, Three Motion Stories, to each student. Note that story C may be a challenge for some students, because the end of that trip could involve fractions of meters.

Choose one of these three stories. Imagine where the beanbags might fall along the track for that trip. Draw the beanbags on the track. Then fill in the table. If you work with a partner, do two different stories.

While students are working, observe to be sure their tables show the distance accumulating, rather than the distance between beanbags. Ask students to explain how their table shows each part of the story. Use this activity as a checkpoint to see how well students understand using tables of positions at regular time intervals to show speed.

- Do students make total distances farther apart to show faster speed and closer together to show slower speed?
- Do they make the total distance remain the same to show a stop?

After completing Student Sheet 5, each student makes a large-size copy of the table on a unlined sheet (it can be the back of used paper). They use dark ink and make the table large enough to fill the whole page, so the numbers can be seen from a distance. Then they put away Student Sheet 5 to use in the next session. Remind students to put their name on both the student sheet and their large-size table.

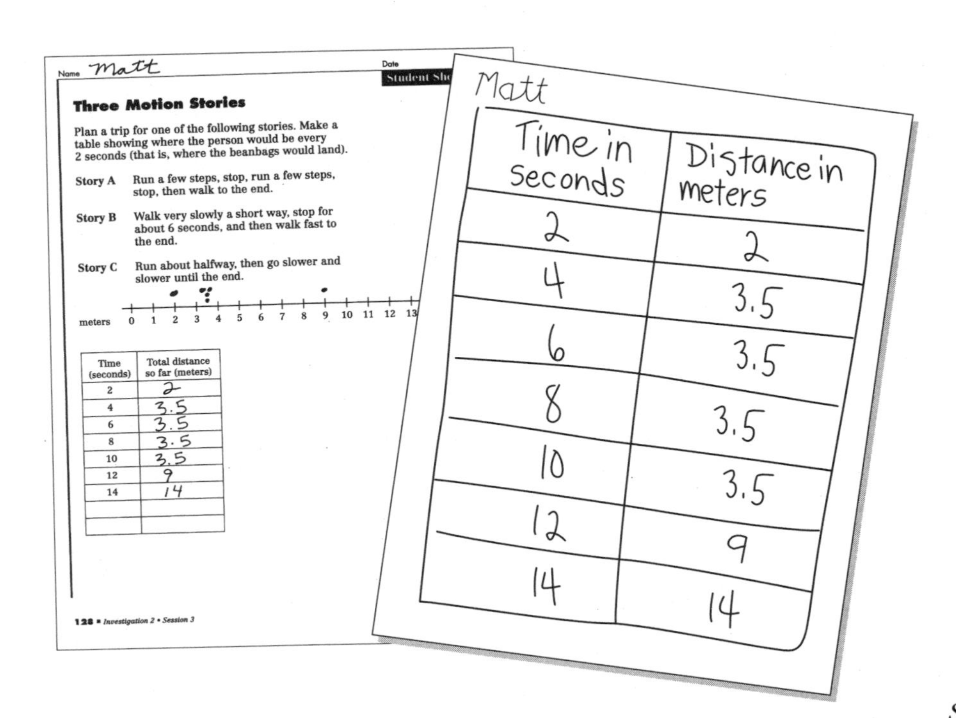

Name Matt Date

Student Sh

Three Motion Stories

Plan a trip for one of the following stories. Make a table showing where the person would be every 2 seconds (that is, where the beanbags would land).

Story A Run a few steps, stop, run a few steps, stop, then walk to the end.

Story B Walk very slowly a short way, stop for about 6 seconds, and then walk fast to the end.

Story C Run about halfway, then go slower and slower until the end.

meters 0 1 2 3 4 5 6 7 8 9 10 11 12 13

Time (seconds)	Total distance so far (meters)
2	2
4	3.5
6	3.5
8	3.5
10	3.5
12	9
14	14

128 ▪ Investigation 2 • Session 3

Matt

Time in Seconds	Distance in meters
2	2
4	3.5
6	3.5
8	3.5
10	3.5
12	9
14	14

Activity

Matching Tables to Stories

Collect all the large-size tables, mix them up, and hand them out again, making sure that no students get their own or their partner's. Students figure out which story the table they received goes with and post the table under that story.

To interpret the tables, students may find it helpful to make a third column showing the intervals, headed something like "Space between beanbags" or "distance in last 2 seconds." They can compute the intervals by subtracting each distance on the table from the next. Some students may want to place beanbags along a track or sketch a track with beanbags to help them visualize the trip.

Check to see if students are able to interpret the intervals between distances to describe portions of the trip as fast, slow, or stopped.

Activity

Similarities Among Tables

When all the tables are posted, students gather around the displays to see what the tables for each story have in common. If students are not sure that a table is placed correctly, or if they think part of a table doesn't fit the story, they can write a question on a stick-on note to place on the table.

When students have had a chance to look at the three groups of tables, they come together for a discussion of similarities in the tables. Start by drawing students' attention to one of the stories and its tables.

How do you know these tables belong with this story? What do they have in common?

Is there any table that does not belong in this group? Why do you think so? What part of the pattern of that story is missing?

Students are most likely to notice the presence or absence of stops, when the space between beanbags is 0 meters (that is, the total distance so far does not change).

Repeat this comparison of the tables for the other two stories in turn. See the **Dialogue Box,** Are They Really the Same? (p. 60), for points that students in one class discussed.

In preparation for graphing in the next session, allow time for students to correct their tables as needed to better fit the story they chose.

Session 3 Follow-Up

Students take home Student Sheet 6, Graph of a Trip. They write the story of the trip shown by the graph. They may write notes directly on the graph to tell how the person was moving for each part—fast, slowly, or stopped. They may make a table from the graph if that would be helpful. Students will share their interpretations of the graph at the beginning of the next session.

❖ **Tip for the Linguistically Diverse Classroom** Students can draw pictures or symbols instead of writing notes to describe the speed.

Are They Really the Same?

STORY B	Walk very slowly a short way, stop for about 6 seconds, and then walk fast to the end.

Christine

Time (sec.)	Total Distance(m)
2	2
4	3.5
6	3.5
8	3.5
10	3.5
12	9
14	14

Desiree

Time (sec.)	Distance (m)
2	1
4	2
6	3
8	4
10	4
12	4
14	4
16	6
18	9
20	11
22	13
23	14

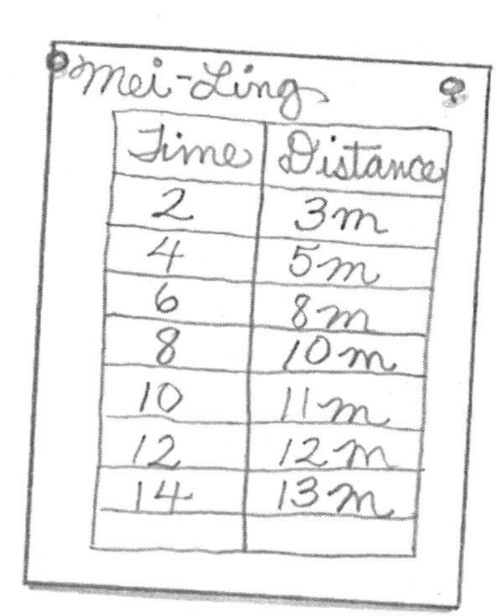

Mei-Ling

Time	Distance
2	3m
4	5m
6	8m
8	10m
10	11m
12	12m
14	13m

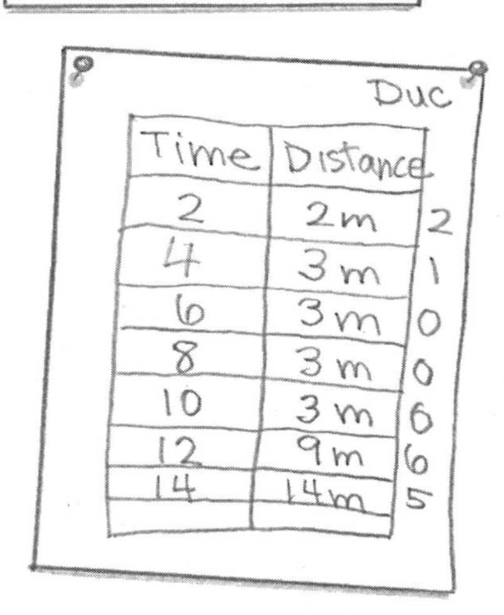

Duc

Time	Distance	
2	2m	2
4	3m	1
6	3m	0
8	3m	0
10	3m	6
12	9m	6
14	14m	5

These students are considering the placement of tables to represent the three motion stories displayed on the wall. They are grappling with an important issue: When do two tables tell the same story, even though they look different from each other? When are two tables really different?

Danny: I don't think Mei-Ling's table goes with story B. It says walk very slowly for a short way, then it says stop for 6 seconds. But she keeps going.

Then what would you have put here, here, and here instead? *[The teacher points to the distances 5 m, 8 m, and 10 m on Mei-Ling's table.]*

Danny: Threes.

Mei-Ling: But I didn't make my table for story B!

OK, then this table tells a different story. Who knows which story it belongs with?

Rachel: I think it's story C: "Run about halfway, then go slower and slower until the end." But yours doesn't really go slower and slower at the end. It's the same speed.

Mei-Ling: That's right. I meant story C. I'll have to fix it.

Take a look at Christine's, Duc's, and Desiree's tables. They're all with story B. How do the tables compare to each other?

Christine: Me and Duc's are the same. We were in different parts of the room, but somehow it got the same.

How does Desiree's table differ? What does she point out?

Duc: There's more time. She figured that to get to 14 meters she needed 23 seconds. We only took 14 seconds.

Desiree's table is a bit different. But it has similarities to Christine's and Duc's too. Desiree has the person stopped for four chunks of time, just as Christine and Duc do. But what part of the story did Desiree take very seriously?

Danny: The first 6 seconds.

Zach: It has her going only 1 meter in 2 seconds, and theirs show 2 meters for the same time.

So they went a little farther when they were walking slowly.

Christine: I think it's the part where she walked fast. She went 6, 9, 11, 13, 14, and those are not sort of far apart like me and Duc's. So I think she made it more seconds.

Duc: Her trip took a longer time. Her going slow was really slow.

Desiree: It says "very slowly." I was trying to show that.

Do you agree that all three tables are ways of telling this same story with slow, stop, then faster?

Christine: Yes, but Desiree's slow is really slow, and our fast is faster than hers.

Session 4

Graphs for Tables

Materials

- Transparency of Student Sheet 6
- Overhead projector
- Completed Student Sheet 5 or enlarged tables (from previous session)
- Student Sheet 7 (1 per student and some extras)
- Stick-on notes (available)
- Centimeter graph paper (1 per student, homework, optional)

What Happens

Working from the table they made in Session 3, students make a line graph of distance versus time. As they did with the tables, they exchange graphs and decide which motion story best fits the graph they receive. Students compare graph shapes that represent the same story to identify common characteristics. Student work focuses on:

- making graphs of total distance from tables of points
- matching graphs to stories and tables
- comparing graphs that fit the same story

Activity

Homework Review: Graph of a Trip

Display the transparency of Student Sheet 6, Graph of a Trip. For a couple of minutes, students explain their interpretations to their neighbors. Then a few students briefly share with the class how they knew when the person on this trip was going fast or slow or had stopped. Encourage students to describe differences in steepness or direction in their own words. Do they see two places where the person goes the same speed? Use this as an introduction to the next activity, in which students make their own graphs.

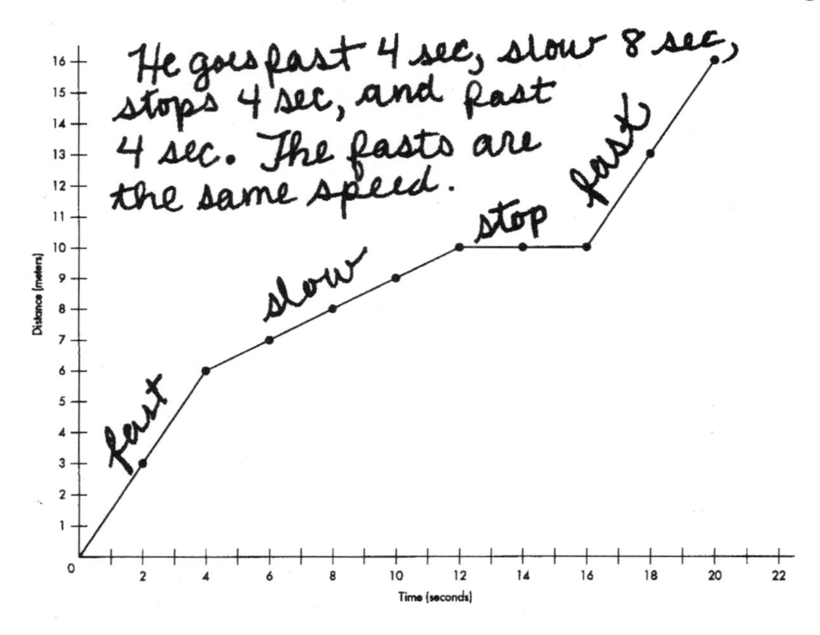

Activity

Making Graphs from Tables

Students will need their work on Student Sheet 5, Three Motion Stories, as they will be making graphs to go with their tables on that sheet. Distribute Student Sheet 7, Graph Template, to each student. Discuss with students that the numbers along the distance axis (vertical) represent accumulated distance, or distance from the beginning of the trip.

Yesterday you chose a motion story and made a table for it. Today you will use this grid to make a graph from your table.

As you observe students working, ask them to explain how they show a slow speed compared to a fast speed, and how they show stopping. Check to see if students who are graphing stories A and B use a horizontal line to represent the stops. Do they understand that to show slowing down, the graph continues to go up, just less steeply? A line going down would mean *going backward.* Look for this also in graphs for story C (run, then go slower and slower). Do students' graphs continue to go up, only changing slope? If they go down when the traveler slows down, take some time to address this misunderstanding.

As students finish their graphs, encourage them to consult with partners to see that their graphs make sense.

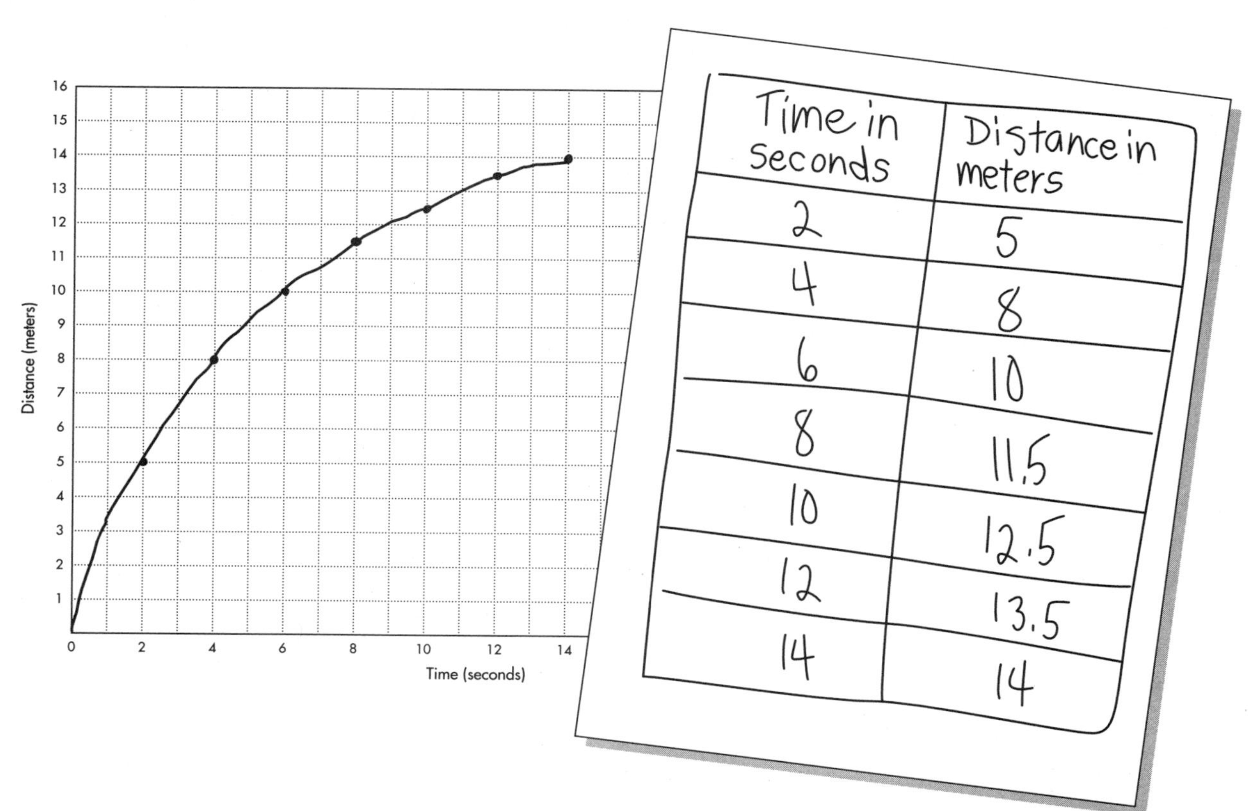

Time in Seconds	Distance in meters
2	5
4	8
6	10
8	11.5
10	12.5
12	13.5
14	14

Activity

Interpreting Graphs

As you did for the tables, collect all the graphs students have made to go with the stories, mix them up, and hand them out again so that nobody gets back his or her own graph. Students post the graphs under the stories they believe they go with. Allow time for students to look at the posted graphs and to comment on them (using stick-on notes).

When students have looked over the graphs, bring them together for a discussion. Find out who thinks their own graph was misplaced. Talk with the whole class about these graphs and others that are controversial. Look for similarity of shape among graphs that go with the same story.

How can you tell when the person was going fastest? going slowest? stopped?

Session 4 Follow-Up

Homework

Students plan a story of any sort of trip—walking, sailing, train, car, or other ideas they have. They write the story, then make a graph to go with it and explain how the graph shows the changes of speed that happen in the story. If they like, they might add a table to make the description of the trip clearer.

The goal is to make the story, graph, and explanation clear enough that they could be used together to teach a fourth grade student how to read the graph. Suggest that students draw axes and write "time" along the horizontal axis and "distance" along the vertical axis before they put their paper away to take home. You might provide centimeter graph paper to those who would prefer to use it.

❖ **Tip for the Linguistically Diverse Classroom** Let students know that they can write the story of their trip in their primary language, supplementing it with the symbols they have developed for varying speeds.

Extension

Change Over Time Students look in newspapers and other printed matter to find graphs or tables that show change over time and bring them in for display and for discussion. They might find some in foreign language newspapers and see if they can interpret them, even if they cannot read the foreign language text. Make a bulletin board of these graphs and tables. If some students bring in graphs that are not about change, designate a separate area of the bulletin board for those graphs and ask students to decide which graphs go in which area.

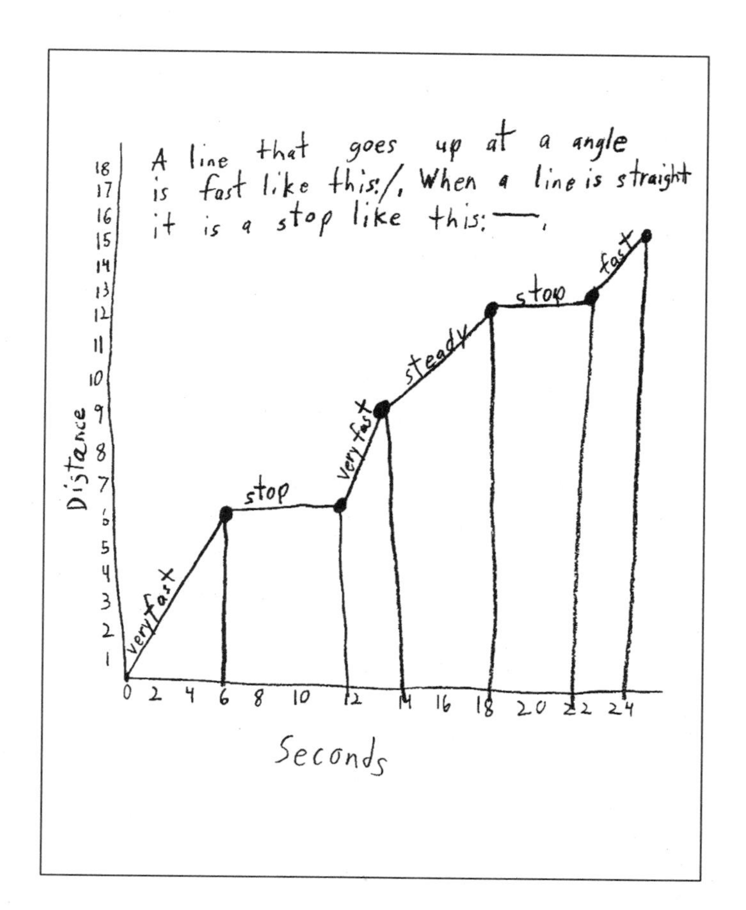

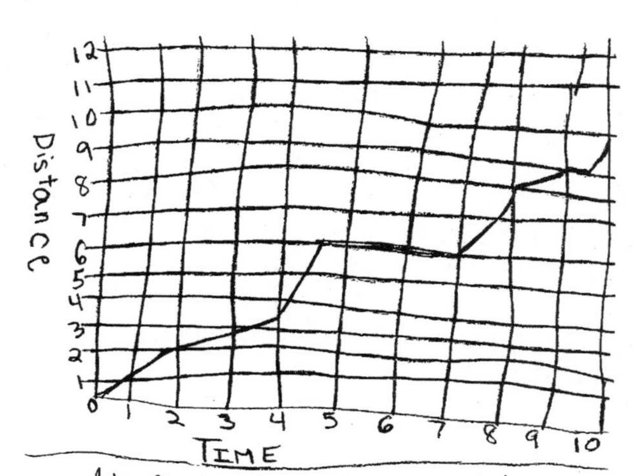

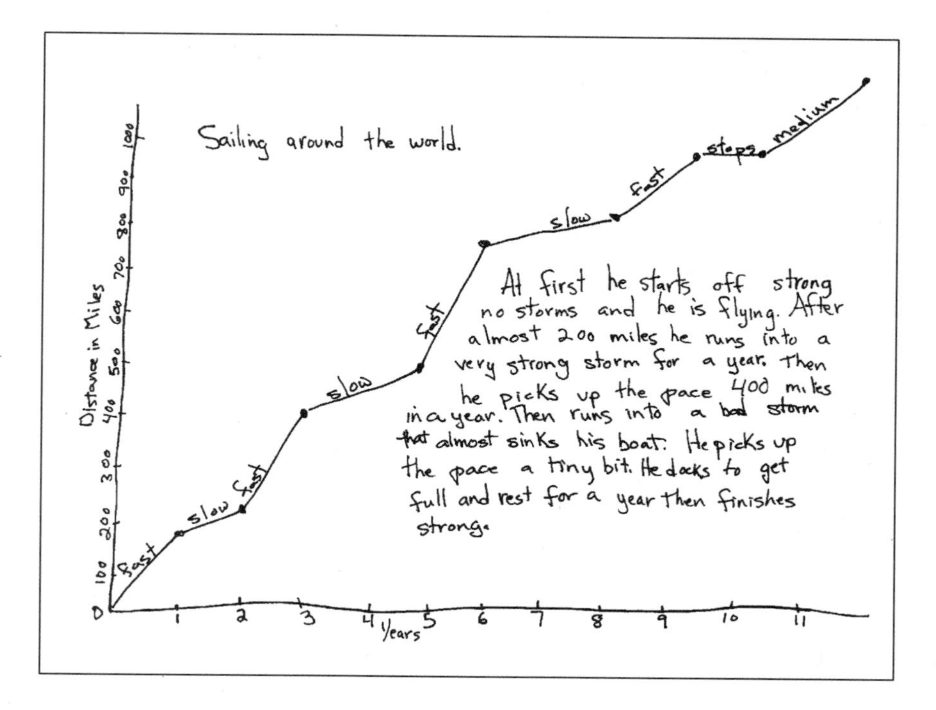

Session 4 Homework These students are explaining how to read graphs showing changes of speed. In the third example, the student's story is communicated appropriately by the graph. Despite his unrealistic choice of numbers (miles covered in a year), it is clear that this student can correctly use the conventions of graphing changes over time.

Session 5

Stories, Tables, and Graphs

Materials

- Student Sheet 8 (1 per student)
- Unlined paper (3 sheets per student)
- Scissors (1 per pair)
- Tape or glue sticks (1 per pair)

What Happens

Students match motion stories, tables, and graph shapes that describe the same trip. They make a table to go with one of the graphs and write a motion story to go with another. Student work focuses on:

- relating tables and graphs of distance traveled to stories of trips
- creating a table of accumulated distance to fit a given graph and story
- writing a motion story to fit a given table and graphs

Activity

Assessment

Matching Stories, Tables, and Graphs

Hand out the two pages of Student Sheet 8, Matching Stories, Tables, and Graphs. Also make available scissors, tape or glue sticks, and unlined paper (the back of used paper is fine).

I have a puzzle for you. On these pages, you'll see three graphs, but only two stories and two tables. One graph has both a table and a story to go with it. One has only a table. One has only a story. The puzzle is to figure out how the stories, graphs, and tables go together. Then finish table A and table B, fill out table C to fit the graph that needs a table, and write a story for the graph that needs a story.

Students cut apart the graphs, stories, and tables. They find the elements that go together and tape or glue them down in groups on the plain paper, then complete the missing parts, and fill in the table for the story that does not have one. The missing story, which goes with graph △ and table B, should be something like this: "Walk about halfway, then turn around and walk back." Table C goes with graph ○ and story 2.

Note: If you think this will be too difficult for your students to do individually, they may work in pairs or small groups and confer.

Assess individual students during the activity or after they have finished by asking them to explain how they decided a group of representations fits together. Ask students to interpret some of the tables and graphs provided by telling the story of the changing speeds.

Following are some things to look for:

- Do students look at intervals between table entries to determine speed? Do they recognize that covering larger distances in 2 seconds means greater speed? Do they show this on the table they invent, as well as when they interpret the tables provided?
- Do students recognize steeper lines on distance vs. time graphs as representing greater speed? Do they recognize a straight line as showing steady speed? Do they recognize a horizontal line as showing zero speed, or being stopped?

Session 5 Follow-Up

Before going on to the next investigation, students might collect all their work from Investigations 1 and 2, including Student Sheets 1–8 and the graph they made of a trip, and clip them together. They could use these materials to explain to an adult friend or family member what they have done in these investigations.

In Investigation 3, students will begin using the *Trips* software on computers. If there is time available before the next session, invite students to explore Setting 1 on the software, trying different starting points and step sizes for the two figures to see what happens when they run their trips.

INVESTIGATION 3

Computer Trips on Two Tracks

What Happens

Session 1: Ways of Making Trips Students discuss how we can walk faster and slower when the number of steps per second stays the same. They look at two ways of making and recording trips: marking successive "steps" along a meterstick on a paper track, and using *Trips*—a computer program with which students can vary the speed of a boy and a girl moving along parallel tracks.

Session 2: Trips on Two Tracks Students explore what happens when they change the speed and the start positions of two travelers along parallel tracks. They create graphs, tables, and lists of commands that correspond to one of three given motion stories. Students work alternately with *Trips* on the computer and with trips along metersticks.

Session 3: Different Kinds of Trips Students continue to work with the *Trips* software and with metersticks, exploring two new types of trips: one in which they can change the speed of either person at some position on the track, and one in which a person's step size can change at every step. Groups use the new trip rules to enact the same three motion stories they explored in the previous session.

Session 4: More Match-Ups Students match each other's tables and graphs (from Sessions 2 and 3) to the three motion stories, then as a whole group discuss any questions and observations about the match-ups.

Sessions 5 and 6: Two Types of Graphs Students explore a different kind of graph, step size vs. time, discussing how to interpret different examples. Working in pairs, they prepare a set of distance vs. time and step size vs. time graphs to represent a "mystery" trip for the rest of the class to guess.

Session 7 (Excursion): Animation Students create two animated flip books, one showing a change over time, and another showing two simultaneous changes that take place at different rates. Students discuss the relationship between successive differences and rate of change.

Mathematical Emphasis

- Developing a vocabulary to discuss motion, e.g., *speed, fast, slow, steady, speed up, slow down, rate*
- Representing motion with number tables, graphs, and verbal descriptions
- Exploring the relationships between time, distance, and speed
- Connecting slope in a graph with rate of change
- Comparing relative motions
- Relating number patterns to graph shapes
- Exploring the relationship between graphs of distance vs. time and step size vs. time

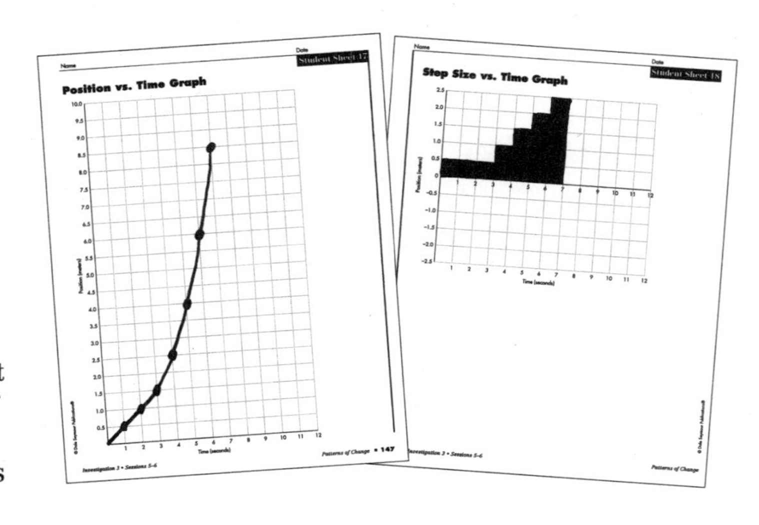

What to Plan Ahead of Time

Materials

- Adding machine tape; see Other Preparation. (Sessions 1–3)
- Metersticks: 1 per group of 2–4 students (Sessions 2–3)
- Masking tape (Sessions 1–3)
- Overhead projector (Sessions 1, 3)
- Colored markers: 2 per group (Session 2)
- Computers—Macintosh II or above, with 4 MB of internal memory (RAM) and Apple System Software 7.0 or later (Sessions 1–3)
- Apple Macintosh disk, *Trips*™ (Sessions 1–3)
- Cuisenaire rods: 1 set of 10 for demonstration (Session 1) and 3–4 sets for student use (Sessions 2–3). Alternatively, use centimeter cubes taped together in trains from 1 to 10 cm long, or narrow strips of tagboard cut in those same centimeter lengths.
- Three large sheets of paper, 2 by 3 feet (Session 2)
- Stick-on notes: 1 pad to share (Session 4)
- Scissors and glue or paste (Sessions 5–6)
- Stick-on notes in light colors for making flip books (2-by-3-inch size works well): 1 pad per student (Session 7, Excursion)

Other Preparation

- Duplicate student sheets (located at the end of the unit) in the following quantities:

  Session 1

  Student Sheet 9, *Trips* Computer Screen (p. 133): 1 per student and 1 transparency

  Session 2

  Student Sheet 10, Trips in Setting 1 (pp. 134–135): 1 per group of 2–4

  Student Sheet 11, Story of a Trip (pp. 136–137): 1 per student, homework. Also make copies as needed for Session 3 homework (1 per student) and a Session 4 extension (1 per student).

  Session 3

  Student Sheet 12, Using the *Trips* Settings (p. 138): 1 per student and 1 transparency

  Student Sheet 13, Trips in Setting 2 (pp. 139–140): 1 per group of 2–4

  Student Sheet 14, Trips in Setting 3 (pp. 141–142): 1 per group of 2–4

  For Sessions 5 and 6

  Student Sheet 15, Two Kinds of Graphs (pp. 143–144): 1 per student

  Student Sheet 16, Mystery Walks (pp. 145–146): 1 per pair plus some extras

  Student Sheet 17, Position vs. Time Graph (p. 147): 1 per pair and extras (optional) for homework

  Student Sheet 18, Step Size vs. Time Graph (p. 148): 1 per pair and extras (optional) for homework

  Student Sheet 19, What's the Story? (pp. 149–150): 1 per student

- Cut the adding machine tape into 110-cm lengths, making 2–3 demonstration strips and 6 strips per group of 2–4 students. (Sessions 1–3)

Continued on next page

What to Plan Ahead of Time (cont.)

- To set up a demonstration trips track, lay a 110-cm length of adding machine tape on a table or the floor where everyone can gather around. Center a meterstick on the strip and lightly tape it in place. Have additional paper strips available. (Session 1)
- Each of the Cuisenaire rods, cube trains, or tagboard strips you will be using to run trips along the metersticks need to be marked at one end to indicate the front. (Sessions 2–3)

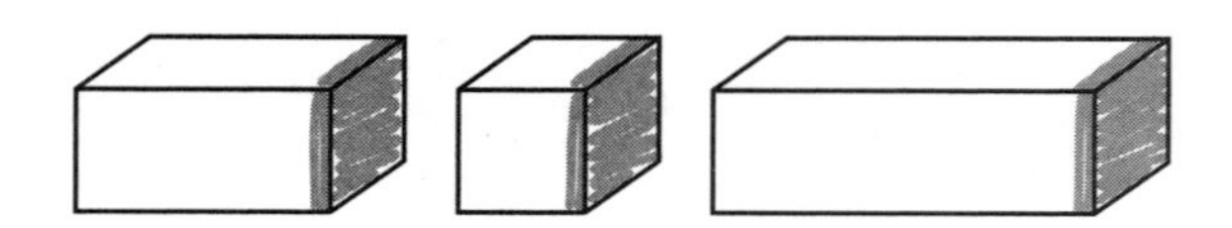

- On the three large sheets of paper, write in large print the three motion stories found on p. 81. (Sessions 2 and 4)
- Cut apart one copy of Student Sheet 16 so that each Mystery Walk is on a separate strip. Each pair needs one story, so for a class larger than 30, duplicate some of the strips. (Sessions 5–6)
- To make a demonstration flip book, use about 20 pages of a 2" by 3" pad of stick-on notes. Show something changing, such as a flower getting taller and taller, or a person walking from left to right across the page. Starting with the last page and working forward is a good strategy. (Session 7)

Computer Preparation

- Use the disk for *Patterns of Change* to install the *Trips* software on each available computer. See p. 111.
- Read the **Teacher Note,** About the *Trips* Software (p. 78). Spend some time experimenting with the program and its three different settings before presenting it to your class. Familiarize yourself with the options available. As needed, refer to the Appendix, Computer Help for *Trips* (p. 111), for further information.

Session 1

Ways of Making Trips

What Happens

Students discuss how we can walk faster and slower when the number of steps per second stays the same. They look at two ways of making and recording trips: marking successive "steps" along a meterstick on a paper track, and using *Trips*—a computer program with which students can vary the speed of a boy and a girl moving along parallel tracks. Student work focuses on:

- predicting trip outcomes based on step size
- predicting how step size and starting position affect trip outcomes

Materials

- Student Sheet 9 (1 per student plus 1 transparency)
- Overhead projector
- 110-cm strips of adding machine tape and meterstick for demonstration
- Masking tape
- 1-, 2-, 3-, 5-, and 8-cm Cuisenaire rods for demonstration
- Computers with *Trips* software installed

Ten-Minute Math: Graph Stories During Investigation 3, do the Graph Stories activity a few times for Ten-Minute Math.

Display a graph on the overhead (see p. 152, or use one that you or a student has made). What story could this graph be telling? Students talk with a partner about possible stories that fit the graph and generate a story to share with the class. They identify the variable shown and tell how it changes. For example:

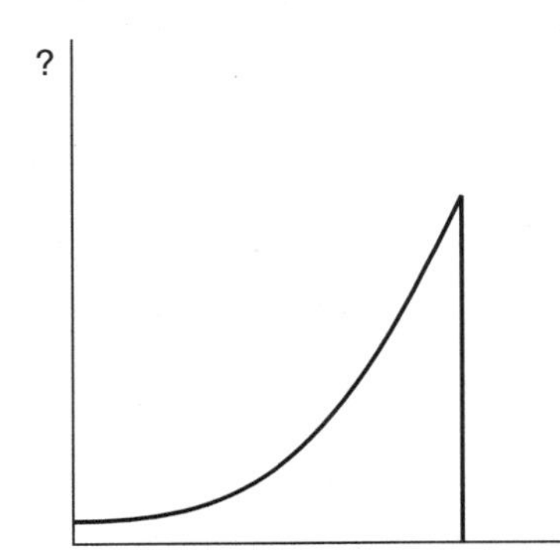

"The balloon was quickly blowing up bigger and bigger and suddenly it popped."

"The weed was growing taller and taller and then I pulled it and it stopped."

If students are having difficulty, offer some guidance: "What could be changing as time goes by? What might be growing or shrinking? going faster or slower? becoming more or less?"

Ask a few students to share their stories, explaining or showing how the story fits the shape of the graph and what is changing with time.

For complete instructions to this activity and variations on it, see p. 108.

Activity

Showing Speed with Step Size

Gather students around the masking-tape track left on the floor from Investigation 2. Recall the beanbag trips in that investigation.

When you took trips along these tracks before, you went at different rates of speed. Sometimes you took 1 or fewer steps per second, and sometimes you took many steps per second. This time we're going to take exactly 1 step every 1 second. Who wants to demonstrate with me?

You and a student volunteer walk side by side, on either side of the tape track. Using a timer, call out "step" every 1 second. Match your step size to the student's so that you both reach the end of the track at the same time. Ask another student to keep track of the number of steps you take.

How many steps did we take to get to the end? Did we both take the same number? With both of us walking at the same rate, 1 step per second, is there any way one of us could go faster than the other? At 1 step per second, how could Leon get to the end of the track before me?

If no one suggests that the student could take *longer* steps, whisper to your student partner to do just that while you take the same size steps as before. Do the trip again, saying "step" every 1 second. Stop when the student reaches the end of the track.

Why did Leon get to the end before me that time? If I wanted to be sure that *I* reached the end first, how would I tell Leon to walk?

Students should understand that when the frequency of the steps remains the same (1 per second), the only way to show slower or faster speed is to change the step size: longer steps for faster speed, and shorter steps for slower speed.

For all the trips in this investigation, we will always use 1 step per second, and only the step size will change.

Note: Although this 1-step-per-second pace may seem artificial to students, we want to establish such a pace in this activity to simplify the way that speed can vary on a walking trip. Once we decide we will always take 1 step per second, then the only way to change our speed is to change our step size. This is the way the *Trips* program works, so this activity helps prepare the students for their work with trips on the computer.

Activity

Trips Along a Meterstick

Gather the class around your demonstration paper trips track, with the meterstick centered on top of the 110-cm length of adding machine tape. Select a 5-cm and an 8-cm Cuisenaire rod (or cube train, or tagboard strip). Be sure you have marked one end of the rods to indicate the "front end." Place one rod on each side of the meterstick track.

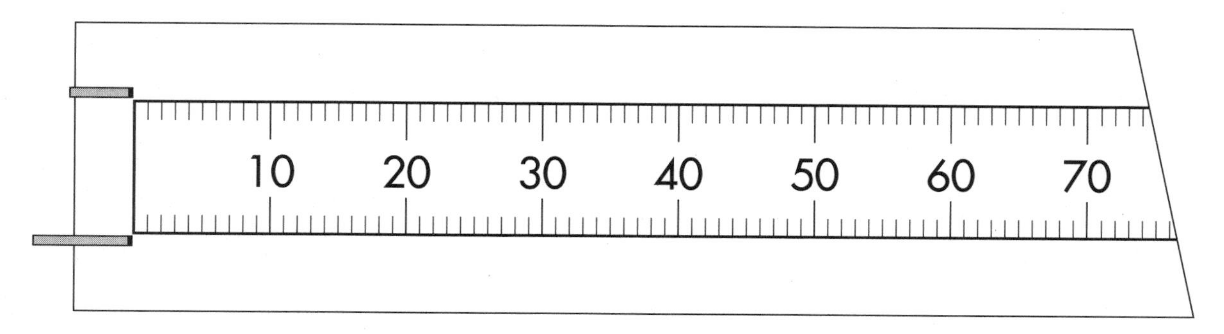

Explain that the rods represent two people taking trips. Each rod takes a step equal to its length. At the start of a trip, the front of the rods would be placed just at the 0 end of the meterstick. Mark both starting positions on the paper tape. When both rods have taken one step, they will be alongside the meterstick.

Demonstrate how students will show trips along the meterstick by moving each rod one length (one step) for each "second." The shorter rod represents someone taking small steps and the longer one represents someone taking large steps. To move the rods, slide them along both sides of the meterstick to their next position, rather than turning them end-over-end or picking them up. Also, call out "Step" at regular intervals to signal the time to move both rods forward one step, simultaneously. Allow enough time between "steps" to mark the new locations of the rods.

Students will run trips the same way, with one person in the group taking responsibility for calling out "Step" to indicate when both rods should be moved. Otherwise, there may be problems with the two students moving the two rods at different rates. After each step, students pause to record the position of the front of the rod with a mark on the paper tape, using a different color for each rod.

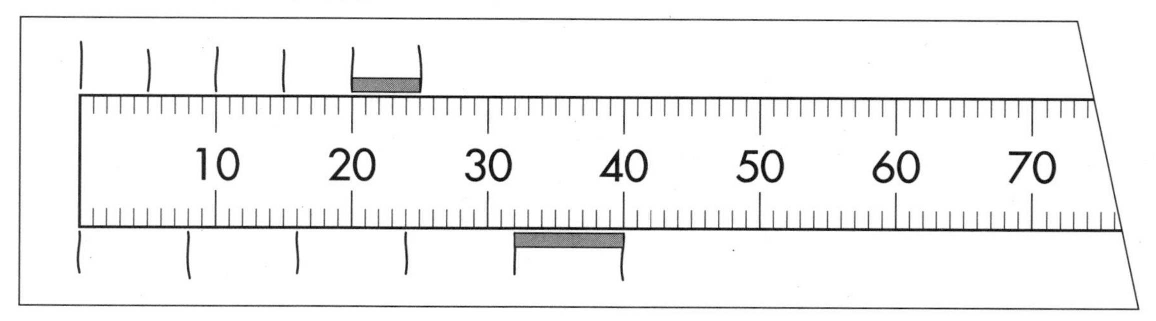

Meterstick, rods and paper tape after 5 "steps"

After taking several steps, ask students to describe how this trip relates to the walking trips you took with a student. Also ask them to describe any patterns in the marks you have made on the paper tape. Some of the patterns they may notice include the regular spacing of the marks on both sides of the meterstick, the difference in the size of the spacing of marks on the two sides, and the way marks on the two sides of the meterstick may sometimes coincide. Students may see other patterns as well.

What will happen as the trip continues?

To mark a new pair of trips, remove the meterstick from the paper demonstration strip and turn the strip over (or use another strip). Tape the meterstick, centered, on the clean strip. Now demonstrate another kind of trip, in which one of the rods does not start at the beginning of the meterstick, but partway along it. The trip that starts later will have marks only from its starting place.

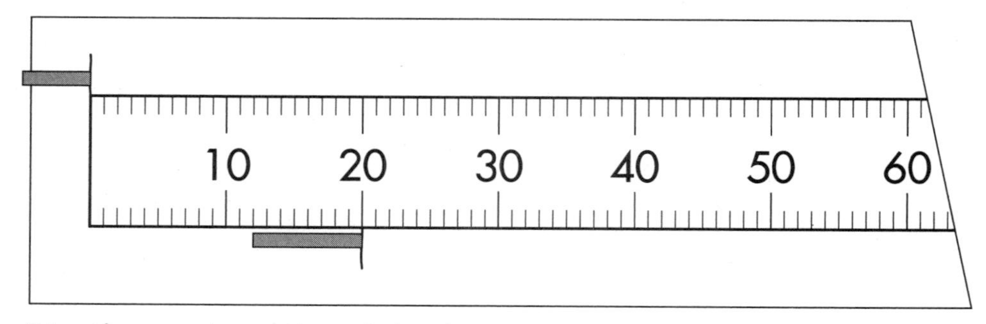

Trip with one starting position at 0, the other at 20

In the next session, you will be planning trips both on and off the computer. When you are working off the computer, you'll use a meterstick on a strip of paper, just as I've shown you. You'll put a rod on either side of the meterstick. Two students will move the rods, sliding them along the stick as one of them calls out "Step." They must make sure to take their steps at the same time, so that both are taking the same number of steps per "second." After each step, you'll mark the rod's progress on the paper strip.

As necessary, reassure the students that it doesn't matter if they take longer than 1 real second to move the rods and make the marks, as long as they keep moving both rods together at the same time. In other words, both rods take exactly the same number of steps in the same amount of time. Remind them to keep the marked end of each rod at the front, showing them where to mark the rod's progress after each step.

Activity

Trips on the Computer

Note: Be sure you have read the **Teacher Note,** About the *Trips* Software (p. 78), and have experimented with the program yourself before presenting it to the students.

To introduce the computer program *Trips* to the class, distribute copies of Student Sheet 9, *Trips* Computer Screen, and display the transparency of this sheet on the overhead. This illustration reproduces the screen as it will open up in Setting 1. The title bar in the main window indicates that you are in Setting 1 (Boy & Girl Start & Step); it will say "Untitled" unless you save your work.

Note: On computers with a smaller screen, the windows are rearranged so that the Command, Graph, and Table windows appear side by side below the main window. If your students will be working at small-screen computers, you may want to sketch this configuration on the board. There is no difference in the way the program operates.

On the computer, you are going to work with a program called *Trips*. You'll see two tracks where a boy and a girl walk, taking steps at exactly the same time, 1 step per second.

You'll notice that time runs fast in *Trips*, so 1 second for the boy and the girl is much shorter than it is for us in real time. This makes it easier for us to watch, because it all happens faster. But for the purpose of the tables and graphs, we will pretend that both the boy and girl are walking exactly 1 step per second.

Gather students once again at your demonstration paper trips track. Center and tape the meterstick on another clean strip of paper tape. Direct students' attention to the Command window on Student Sheet 9.

The words and numbers in this window are commands. They are setting up a trip for the boy and girl, telling them how to move along the tracks from the house to the tree. Let's talk about how this trip will go.

```
startboyposition 0
startboystep 2
startgirlposition 10
startgirlstep 1
```

What do you think the girl and boy would do to follow the instructions in the Command window?

As students give their ideas, demonstrate with the meterstick and paper track, using 1-cm and 2-cm Cuisenaire rods for the girl's and boy's steps. According to the commands, the boy starts at position 0 and walks with a step size of 2. The girl starts *not* at 0, but at 10 on the track. Her step size is 1.

Show this by placing the 2-cm rod (for the boy) on one side of the meterstick, with the marked end of the rod at 0. Place the 1-cm rod (for the girl) on the other side, with the marked end of her rod at 10 cm. Mark each starting position on the paper tape, and mark a few steps as you move each rod to demonstrate the trip.

Next, show how to set up a new trip by changing commands. Use the transparency of Student Sheet 9 to demonstrate.

We can change the step size and the starting position of both the boy and the girl just by changing the numbers in the commands. We put the cursor at the end, like this *[with an overhead pen, draw a line to show a cursor at the end of the first command]:*

```
startboyposition 0|
```

Then we delete the number and type a new one. Suppose that I change it like this. *[Mark out the 0 and write 50. Also mark out and change the other commands on the transparency as follows.]*

```
startboyposition 50
startboystep 1
startgirlposition 0
startgirlstep 3
```

What will the boy and the girl do now? Who will get to the tree first with this set of commands?

Set up 1-cm and 3-cm Cuisenaire rods on the meterstick according to these commands: The 1-cm rod for the boy (small steps) starts at 50, and the 3-cm rod for the girl (larger steps) starts at 0. Ask the class to count seconds as you take steps along the meterstick track with the two rods. Mark the steps as you move the rods, using a different color pen for each rod. After a few steps, ask:

Who do you think will get to the end first?

Volunteers move the rods through the rest of the trip to see whose prediction is correct. As necessary, remind them that the boy and girl take their steps at exactly the same time.

Could you tell me the story of that trip, from your point of view, as an observer?

One possible response would be: The girl is going really fast. She catches up to and passes the boy, who is going slow.

Now let's think about the story of the same trip from the point of view of the *girl*. If I were the girl, I might tell the story of the trip this way:

I started out way behind the boy, who was already halfway to the end by the time I got going. So I went really fast and caught up to him more and more. Finally, at 75, I passed him. Then I kept going really fast and got to the end first.

Ask for a volunteer to retell the story once again, this time from the point of view of the *boy*. Make sure that students understand the idea of talking as if they were the boy—what *he* sees and what *he* experiences—rather than speaking as an observer. In later activities, students will be asked to tell the story of their trip from the point of view of the girl or the boy.

Activity

On Computer

Exploring the *Trips* Software

Students spend the rest of Session 1 getting acquainted with the *Trips* software. Two to three pairs of students gather around a computer and take turns running the trips. The first pair plans a trip for the boy and girl by filling in different values in the Command window. Before starting any trip, all the students at the computer guess who will get to the end of the track first. After all guesses are in, students press the Go button to start the trip.

When one pair of students has planned and run two trips, another pair takes control of the computer, and the first pair joins in guessing and watching. Students continue with this rotation until the end of class.

The goals of this first interaction are to learn how to run trips on the computer and how to change the *step size* and *start positions*. Suggest that students sometimes have the boy get to the end first, and sometimes the girl. They could also try for a tie with the boy and girl starting at different places.

Point out one similarity with the beanbag trips students walked themselves in Investigation 2: As the boy and girl walk at varying speeds, a little arrow appears under their feet every 2 "seconds," as if a beanbag had been dropped at that point. The students can use the pattern of little arrows to understand what happened during their trip, just as they used the patterns of beanbags left on the floor to recreate and understand their walking trips.

Teacher Note — *About the Trips Software*

Trips is a computer program designed for mathematical exploration of motion. In the *Trips* activities, students control the movement of two people simulated on the computer screen, a boy and a girl. They walk along parallel tracks, numbered from 0 to 100, between a house and a tree. Their travel along these tracks is termed a "trip."

Students can define the initial position, direction, and speed of both figures with commands such as " startgirlposition 10" (meaning that the girl will start her walk at position 10 on her track). To make the girl or boy go faster or slower for a particular trip, students change the size of their steps.

Students see the data for any trip displayed in several ways: along the number-line tracks, in the table that records the boy's and the girl's changing position over time, and in a graph showing either position vs. time or step size vs. time for both walkers. These representations are connected so that when one is manipulated, changes in the others can be observed.

Using *Trips,* students investigate relationships involving change and motion: What happens if we change the speed of the girl? What happens if we change the starting position of the boy? What if we change the girl's speed part way through the trip? What if we change direction?

Starting Up Trips To open the *Trips* program, double-click on the *Trips* icon and wait for the title screen.

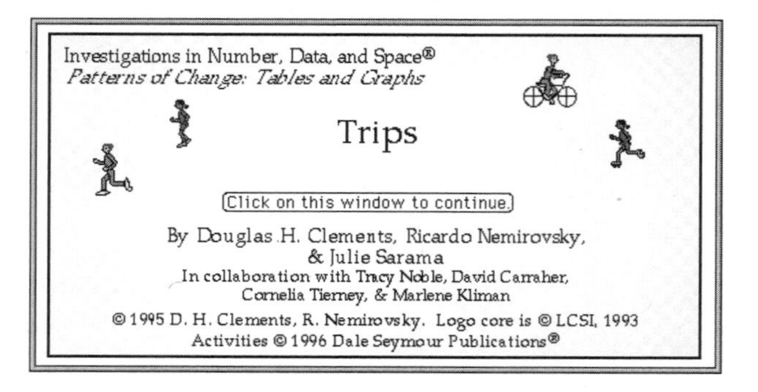

After a single click on this screen, the four windows of the *Trips* program will open.

Running a Trip Students have a choice of three "settings" that run different kinds of trips (further described in the **Teacher Note,** Using the Three *Trips* Settings, p. 85). *Trips* always opens in Setting 1, which is the first setting students use for their work in this unit. The illustration on p. 79 shows the four windows as they appear in Setting 1. (If you have a computer with a small screen, you will see the same four windows in a slightly different arrangement.) When *Trips* first opens, a boy and a girl are shown at position 0 on the two numbered tracks.

You will see that a particular trip has already been described in the Command window. Click on the **Set** button to move the boy and girl to their starting positions. Run the trip by clicking on the **Go** button.

When the trip starts, the Go button toggles into a Stop button. If you want to stop the boy and girl while they are moving, click on this button.

When a trip is running, the elapsed time and the positions and step sizes of the boy and the girl are displayed on the screen. For the purposes of the activities, we say time on these trips is measured in "seconds." However, to speed things up, these are simulated seconds that run many times faster than actual time.

The **Step** button lets you "step through" a trip to observe it one step at a time. Click on the Step button to start a trip if you want to step through the trip from the start, or click on it *during* a trip to step the remainder of the trip from that point on. Each subsequent click will advance the time by 1 "second." While you are stepping, you can click outside the main window to resume the trip, or click on the Stop button to stop it.

Continued on next page

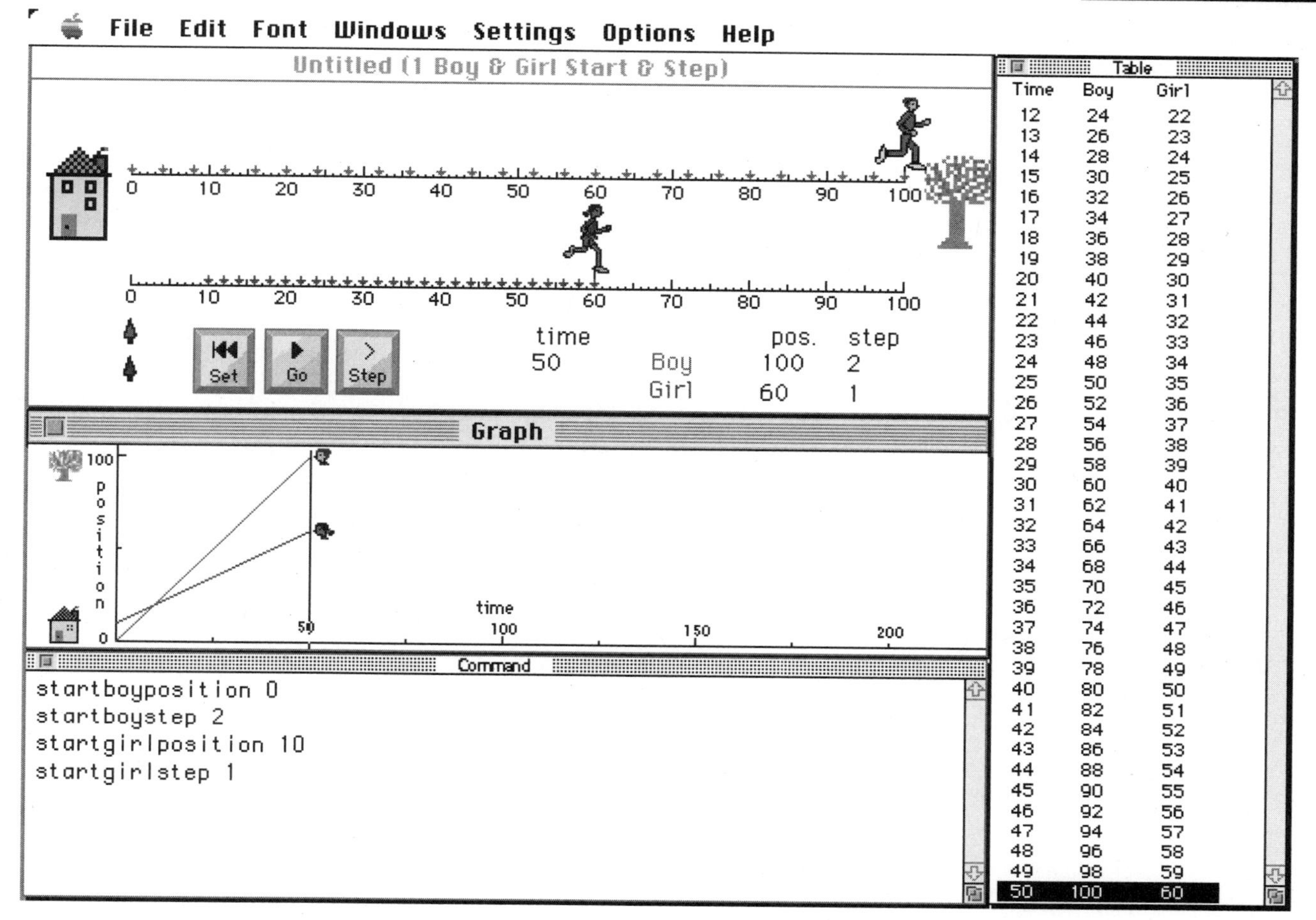

Editing the Commands For each setting, the basic commands are provided in the Command window. The students' task is to change the inputs to these commands (that is, the numbers) to create a certain kind of trip. For example, in the trip that is described when you open in Setting 1, the boy reaches the tree before the girl does. Students are asked to change the inputs to the commands to create a trip in which the girl gets to the tree way ahead of the boy.

To change the inputs to the commands, students can simply click directly to the right of the number, use the **<delete>** key to erase the number, and then type a new one.

Analyzing Your Trips After running a trip, you can click on the Graph or Table window to see a graph or table of that trip. The graph is initially set to show position vs. time. To graph step size instead of position, select **Graph Step** from the **Options** menu. If you wish to see the graph or table built during the trip, as the boy and girl are moving along the tracks, select **Graph During Trip** or **Table During Trip** from the **Options** menu.

After a trip, you can use the Graph or Table window to examine individual points in that trip, looking for correspondences between points in the graph or the table and the various positions of the boy and girl. Click on a position on the graph or on a row in the table (you can also use the arrow keys to select rows in the table). The boy and girl will return to the positions they were in at that time, and the graph and table will show the data for that time: the graph with a vertical line, and the table with a highlighted row.

Continued on next page

Teacher Note

continued

For a visual record of each trip similar to that provided by the beanbags dropped in classroom activities, students can mark the positions of the boy and girl at regular intervals with little arrows (termed **Marks** in the **Options** menu) just below the numbered track. In the initial set-up, arrows appear every 2 steps. You can change the interval between them by typing in the Command window `marksevery`, space, and the interval you want. To get marks at intervals of four steps, for example, type `marksevery 4`. (This would be useful, for example, when the boy and girl are walking very slowly, as with a smaller interval, marks every 2 seconds would fall on top of each other.)

Other Options In the main Trips window, two pointers (red for the boy and blue for the girl) are stored just below position 0 on the tracks.

You can drag these pointers to any position that you wish to mark. These are useful for making predictions before running a trip. For example, students might predict where the boy will be when the girl reaches the tree, or at what point one figure will pass the other.

When you have placed a pointer somewhere along the track, its position appears in the Command window (for example, `boypointer 20`). You can change the position of the pointer either by dragging it where you want it or by changing the numerical input to this command.

Any time you would like to verify the position of a pointer, you can shift-click on it (hold the **<shift>** key down while you point at the object and click the mouse button); the object's name and position on the path are then shown. You can also use this feature to check the position of the boy or the girl when they are stopped along the track.

To hide the boy and girl during a trip (as students may want to do for the extension activity, The Challenge, p. 91), select **Screen** from the **Options** menu. Four square screens will appear, covering the tracks and the figures. Clicking on the individual square screens shrinks them to reveal what's behind; clicking again makes them grow to their original size. To remove them entirely, select **Screen** again from the **Options** window.

For variety, students can change the two "runners" into a biker (the boy) and a skater (the girl) by making the appropriate selection on the **Options** menu. This choice makes no change in how the trips operate.

Other Commands For classes or individual students who want to explore the *Trips* software further, refer to Making Your Own Trip: Commands (p. 114) in the Appendix: Computer Help for *Trips*. This section provides information on other commands students can use to create trips.

Session 2

Trips on Two Tracks

What Happens

Students explore what happens when they change the speed and the start positions of two travelers along parallel tracks. They create graphs, tables, and lists of commands that correspond to one of three given motion stories. Students work alternately with *Trips* on the computer and with trips along metersticks. Student work focuses on:

- understanding the relationship between speed, distance, and time
- combining discrete and continuous descriptions of motion

Motion Story 1

The girl gets to the tree way ahead of the boy.

Motion Story 2

The girl starts behind the boy, but she passes him and gets to the tree frist.

Motion Story 3

The boy starts at the tree and the girl starts at the house. The boy gets to the house before the girl gets to the tree.

Materials

- Display copies of three Motion Stories
- Student Sheet 10 (1 per group)
- Adding machine tape (3 strips per group)
- Metersticks (1 per group), masking tape
- Cuisenaire rods (3–4 sets of 10 to share)
- Markers (2 colors per group)
- Computers with the *Trips* software installed
- Student Sheet 11 (1 per student, homework)

Activity

Working with Motion Stories

We're going to be planning more trips for the boy and girl you saw yesterday in the *Trips* program. Half of you will work on the computer while the rest of you work with the meterstick-and-paper tracks. Then you'll switch places. The goal of this activity, whether you are working on or off the computer, is to have the boy and the girl move according to the three motion stories posted on the wall.

Ask volunteers to read the three stories.

Call attention to story 3, in which the boy starts at the tree at the right end of the track. For most of the trips so far, students will probably have had the boy and girl walking from left to right (from the house to the tree). In order to turn a figure around so it will walk in the opposite direction (from the tree to the house), we have to enter a special command.

If you put a minus sign in front of the number that tells the step size, the computer will make the boy or girl turn around and walk in the other direction, toward the house. On the computer, we can use a hyphen for the minus sign.

As an example, write on the board these four commands:

```
startboyposition 50        startgirlposition 0
startboystep -1            startgirlstep 3
```

How would the boy and the girl walk with these instructions? (The boy starts the trip halfway between the tree and the house and walks toward the house slowly. The girl starts at the house and walks toward the tree, moving faster.)

Put students in groups of 2 to 4, depending on computer availability; half the groups should be able to use the computers at the same time. Distribute a copy of Student Sheet 10 to each group. Designate which groups will work first at the computer.

When students successfully move the boy and the girl in a way that matches one of the motion stories, either at the computer or with the meterstick, they fill in their starting commands on the first page of Student Sheet 10 and go on to the next story. Explain that when they have recorded a set of commands for all three motion stories, they are to choose their favorite story and use it on the second page of Student Sheet 10, filling in the table, making a graph, and writing a description of the trip.

Note: You may want to discuss two features of Student Sheet 10 (page 2): the story description of the trip and the time intervals. Point out that the description they write should tell how the trip would look from the point of view of the boy or the girl. Emphasize that this description should be more specific than the corresponding motion story. For example, a version of story 2 from the boy's point of view might be:

> I started *well* ahead of the girl and went to the tree really slow. She passed me at around 70. I saw her getting to the tree first.

Students will find no numbers beyond 0 in the time column of the table and along the time axis in the graph. This allows them to create their own scale by choosing time intervals, based on the length of the trip they want to show. They will need to choose intervals that that enable them to show the whole trip clearly in their table and graph. For example, Lindsay (student work, p. 83) wanted to show a trip that took 25 seconds. She could not show the whole trip using 1-second intervals, so she chose 2 seconds. Because the last interval in her table is just 1 second, the last data point in her graph falls midway between two vertical grid lines.

A student who is showing a very fast trip—say, one that is over in 6 seconds—may want to number alternate lines on the time axis (two vertical lines equal 1 second), to stretch out the graph. Circulate to help students as needed with the selection of appropriate time intervals when they begin making their tables and graphs.

For Trips with Metersticks As the computer groups begin work, hand out materials for the rest of the class to work with. Each group will need 1 meterstick, paper strips, and masking tape to lightly fasten the meterstick in place on a paper strip. Also make available 4–5 sets of Cuisenaire rods, cube trains, or tagboard strips in 1-cm to 10-cm lengths, each marked to show a "front" end.

Whether you are working at the computer or with the meterstick track, you will fill in the same information on the student sheet. When you use the meterstick and rods, mark your trips along the paper strips on each side of the meterstick, using a different color for each "traveler."

Advise students that there is more than one way of working out these trips, so different groups may get different answers that are both right.

Divide the remaining time in the session in half. Halfway through, students at the computer switch with those at the meterstick tracks.

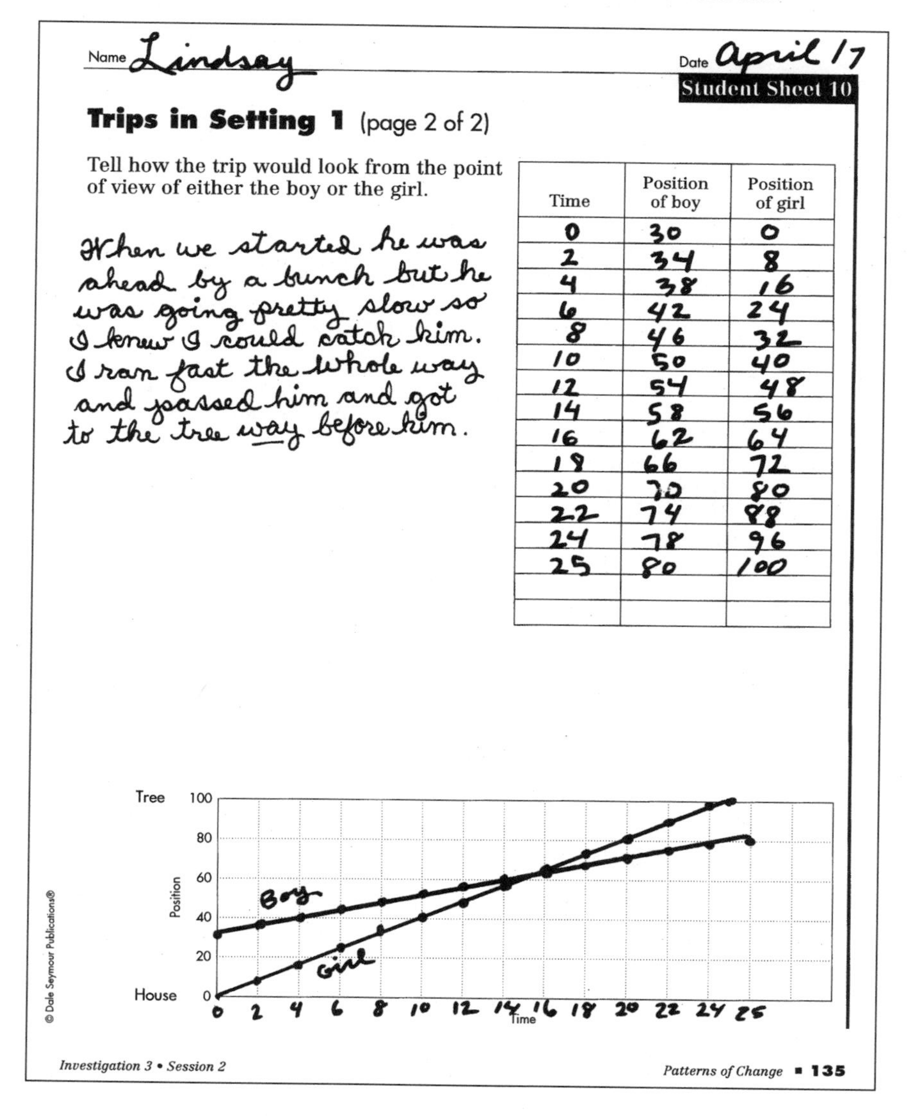

Name Lindsay

Date April 17

Student Sheet 10

Trips in Setting 1 (page 2 of 2)

Tell how the trip would look from the point of view of either the boy or the girl.

When we started he was ahead by a bunch but he was going pretty slow so I knew I could catch him. I ran fast the whole way and passed him and got to the tree way before him.

Time	Position of boy	Position of girl
0	30	0
2	34	8
4	38	16
6	42	24
8	46	32
10	50	40
12	54	48
14	58	56
16	62	64
18	66	72
20	70	80
22	74	88
24	78	96
25	80	100

Tree
House
Position
100
80
60
40
20
0
Boy
Girl
0 2 4 6 8 10 12 14 16 18 20 22 24 25
Time

Investigation 3 • Session 2 *Patterns of Change* ▪ **135**

Teacher Checkpoint

Running and Recording Trips

Observe students as they work. At the end of this session, collect all copies of Student Sheet 10, Trips in Setting 1, along with the paper strips students have used. Use these as a guide to determine the following:

- Are students able to create lists of commands that are appropriate to each story?
- Are students recording data accurately on the tables? Step size is constant for all these trips, so increments in the table should be constant. The numbers in the time column should match the numbers written along the time axis in the graph.
- Can students create graphs for their chosen stories? Are the lines on the graphs straight, reflecting constant step size?
- Are students making good choices of time intervals, so they can fit all their data in the table and graph? Are the intervals small enough that the data fills more than half the table and is not all squeezed into a corner of the graph?

Return the work to the students with comments. Be sure they save their work on Student Sheet 10 for Session 4, when they will be using the tables and graphs in another activity.

Session 2 Follow-Up

Homework

Students take home the two pages of Student Sheet 11, Story of a Trip. They write their own motion story for the girl and boy and fill in the tracks, tables, and graph according to the story.

❖ **Tip for the Linguistically Diverse Classroom** Students with limited English proficiency can illustrate their story, perhaps using frames to show where the girl and boy begin, how their positions change as they race, who reaches the end first, and where the other one is at that point.

Extension

- **Another Motion Story** If some students need a more challenging motion story, propose the following:

  The boy and the girl start together at one place along the track that is *not* at either end or in the middle. The boy walks toward the house and the girl toward the tree. The boy gets to the house at the same time as the girl gets to the tree.

Using the Three *Trips* Settings

Teacher Note

Students begin their work with the *Trips* software in Setting 1. As they continue working through Investigation 3, other activities require them to use Setting 2 and Setting 3. Changing the setting (with the **Settings** menu) changes the set of commands that appears in the Command window, making different types of trips.

Settings
✓1 Boy & Girl Start & Step
2 Change Step at a Position
3 Change Step Constantly

Setting 1: Boy & Girl Start & Step In this setting, you can change the starting position and step size of both the boy and the girl. The setting opens with these commands:

```
startboyposition 0
```
(Boy starts at position 0.)
```
startboystep 2
```
(Boy has step size of 2.)
```
startgirlposition 10
```
(Girl starts at position 10.)
```
startgirlstep 1
```
(Girl has step size of 1.)

Setting 2: Change Step at a Position In Setting 2, you can still change the starting position and step size of both the boy and the girl. In addition, you can change their step size when they reach a certain position on the track. Setting 2 opens with these commands:

```
startboyposition 0
startboystep 1
startgirlposition 0
startgirlstep 2
changeboystepto 4 [when
boyposition = 25]
```
(Change boy's step size to 4 at position 25.)
```
changegirlstepto 3 [when
girlposition = 25]
```
(Change girl's step size to 3 at position 25.)

Setting 3: Change Step Constantly In the third setting, you can again change the starting position and step size of both figures. In addition, you can change their step size constantly, once each second, so they are going increasingly faster throughout the trip. Setting 3 opens with these commands:

```
startboyposition 0
startboystep 0
startgirlposition 0
startgirlstep 1
changeboystepby 2 [always]
```
(Increase the boy's step size by 2 at every step.)
```
changegirlstepby 1 [always]
```
(Increase the girl's step size by 1 at every step.)

Students may find it useful to increase step size by small decimal numbers such as 0.1 or 0.25 so that the figures increase speed slowly. They might also experiment with *decreasing* speed by using negative inputs.

Student Sheet 12, Using the *Trips* Settings, provides examples of the three different types of trips with unfinished tables of the boy's and girl's step size and position changing over time. You can use this sheet in Session 3 to introduce the settings. As students finish the tables and share their results, you can check their understanding of how the three settings work.

For the remainder of Investigation 3, students explore how changing the numerical inputs in the different settings affects the outcome of the trips. They see that the three settings can be used to set up a variety of trips that match the same three stories about the motions of the boy and girl.

"The Boy Is Going to Win..."

Matt and Tai are using the *Trips* software. Before they run a trip, the teacher asks them what they think will happen. These are the commands:

```
startboyposition 0
startboystep 2
startgirlposition 50
startgirlstep 1
```

Tai: The boy will get creamed.

Matt: The boy is going to win, 'cause he's got a step of 2. Win! Win!

Tai: The girl's got to win because she's so far ahead.

Matt *[a few seconds before the trip ends]*: It's going to be a tie.

How did that happen?

Matt: It makes sense. The girl had a halfway of the distance head start, and she's only half the speed of the boy.

Tai: The boy can go twice as fast as the girl.

What is happening here? *[The teacher points to the graph window.]*

Matt: This line *[indicates girl's position]* goes up high right away because she got a head start.

Tai and Matt are figuring out the trade-off between step size and head start. Initially Tai expected that the girl would win because she started much closer to the tree. Matt, on the other hand, expected that the boy would win because of his larger step size. Tai and Matt interpreted the tie as a result of the exact compensation of the girl's head start for her smaller step size.

This kind of mutual compensation between factors—generally speaking between speed, distance, and time—is a central aspect of this unit. Encourage students to think and talk about the ways the factors of a trip interact. For example, they might reason, "If you keep the same number of steps per second, the larger the step size, the faster you go."

Session 3

Different Kinds of Trips

What Happens

Students continue to work with the *Trips* software and with metersticks, exploring two new types of trips: one in which they can change the speed of either person at some position on the track, and one in which a person's step size can change at every step. Groups use the new trip rules to enact the same three motion stories they explored in the previous session. Student work focuses on:

- describing differences between trips with constant step size and trips with changing step size
- interpreting position vs. time graphs with straight, broken, and curved lines

Ten-Minute Math: Graph Stories Continue to present the Graph Stories activity, using graphs from p. 152 or your own. Encourage the class to interpret the graphs both in terms of varying speed on walking trips and in terms of other things that change over time. For complete instructions and variations on this activity, see p. 108.

Materials

- Overhead projector
- Student Sheet 12 (transparency and 1 per student)
- Demonstration paper tape, meterstick, and Cuisenaire rods
- Student Sheet 13 (1 per group)
- Student Sheet 14 (1 per group)
- Adding machine tape (3 strips per group), masking tape
- Metersticks, rods, and markers from Session 1
- Computers with *Trips* installed
- Another copy of Student Sheet 11 (1 per student, homework)

Activity

How Settings 2 and 3 Work

Before you begin, divide the students into two groups: one that will use the computer first and another that will use the metersticks first. The students at the metersticks will work with the type of trip described by Setting 2, and those at the computer should start with the type of trip run in Setting 3. Students who did not have time to complete the activity in Session 2 should continue to use Setting 1.

In all the trips we have done so far, the step size stayed the same throughout the trip for both the girl and the boy. Now we are going to learn some ways of changing the step size along the way.

In the *Trips* program, you have three settings to choose from. Setting 1 is the one that you have been using: step size is set at the beginning of the trip and doesn't change. In Settings 2 and 3, you can change the step size during the trip and thereby change the boy's or girl's speed.

Hand out Student Sheet 12, Using the *Trips* Settings. Show the transparency of this sheet on the overhead projector as you help students understand the three setting options.

At the top of this sheet, you'll see commands for a trip in Setting 1. You used this kind of command yesterday when you did trips at the metersticks and on the computer. The boy's step size for this trip is –1. What does that mean?

Ask a volunteer to explain how the entries in the first table were established. Following student suggestions, fill in one or two more entries. Students then complete this first table before you explain Settings 2 and 3.

In Setting 2, the step sizes of the boy and the girl can be changed when they reach a certain position on the track. In the example on Student Sheet 12:

> The boy starts with step size of 2 and changes it to 8 when he reaches the position 4 on the track.
>
> The girl's step size starts at 4 and changes to 6 when she reaches the position 12 on the track.

Demonstrate the use of the meterstick and rods to represent this trip in Setting 2. Note that you will have to change the size of the rods you are using partway through the trip. Have at hand the 2-cm and 8-cm rods for the boy and the 4-cm and 6-cm rods for the girl.

Using your demonstration meterstick centered over a strip of paper tape, start the trip with the 2-cm and 4-cm rods. As students refer to the table to tell you the step size and position of each walker at each second, slide the rods along the meterstick and make colored marks that correspond to positions of the boy and the girl. Change to the larger rods when students tell you to.

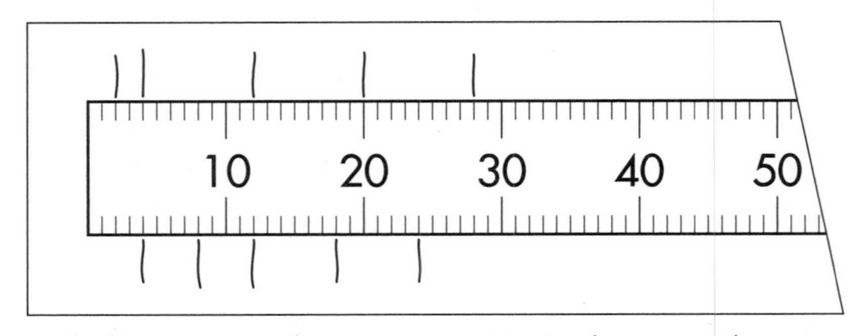

Marks showing step size changes at postion 4 (top) and position 12 (bottom)

Do not continue past step 5; ask students to complete the Setting 2 table on Student Sheet 12 and to share their results and any patterns they see.

In Setting 3, the step size is changed at every step. In the example on Student Sheet 12:

> The boy's step size increases by 1 every step.
>
> The girl's step size *decreases* by 1 every step.

In Setting 3, the boy or the girl can change direction as they walk. If their step size goes down to 0 and then becomes negative, they will slow down, stop, then start again in the opposite direction.

Using the reverse side of your paper strip, demonstrate how you could try using rods and the meterstick to run the Setting 3 trip on Student Sheet 12. Start with no rod (0-cm) for the boy's step and use the next larger rod (1-cm, then 2-cm) for each successive step. At the same time, start with the 10-cm rod for the girl's first step and decrease the size of the rod by 1 for each step. This process will be cumbersome; explain that for this reason, students will be doing trips in Setting 3 only on the computer, not at the metersticks. At the same time, watching you change the rods at each step will help the students visualize what is happening in Setting 3.

Ask students to complete the table for Setting 3 and to share their results.

Activity

Running Trips in Setting 2 and Setting 3

Make available copies of Student Sheet 13, Trips in Setting 2, and Student Sheet 14, Trips in Setting 3, to each group as they go to work on the metersticks (in Setting 2) or the computer (in Setting 3). Students will recognize the three motion stories on both sheets as the same stories they worked with in Setting 1. Here they will discover different trips that match the same stories.

Students work in small groups creating trips for the setting they have been assigned. As in Session 2, students fill in the first page of their student sheets for every motion story. They then choose their favorite story to make a table, draw a graph, and write a more specific account of the trip on the second page.

Half of the groups work on the computers and the other half on the metersticks and paper strips. About halfway through the session, the groups working on the computers with Setting 3 should switch to the metersticks and try Setting 2. The groups who have been at the metersticks should move to the computers. They may first want to check their Setting 2 trips (done on the paper strips) on the computer. If there is time, they should also try at least one trip using Setting 3.

Note: In Setting 2, students might set up a trip in which the step size of the boy or the girl changes at a point that does not coincide with his or her position at any particular second, because the boy or girl *steps over* that position instead of landing on it. If this happens on the computer, students may find the boy or girl taking one step with a step size that is neither the original step size nor the new step size, but a number halfway in between these two. See the **Teacher Note,** Changing the Step Size in Setting 2 (p. 92), for a discussion of how to handle this in your class.

Students need to keep their completed Student Sheets 13 and 14 with their previously completed copies of Student Sheet 10 for use in the next session.

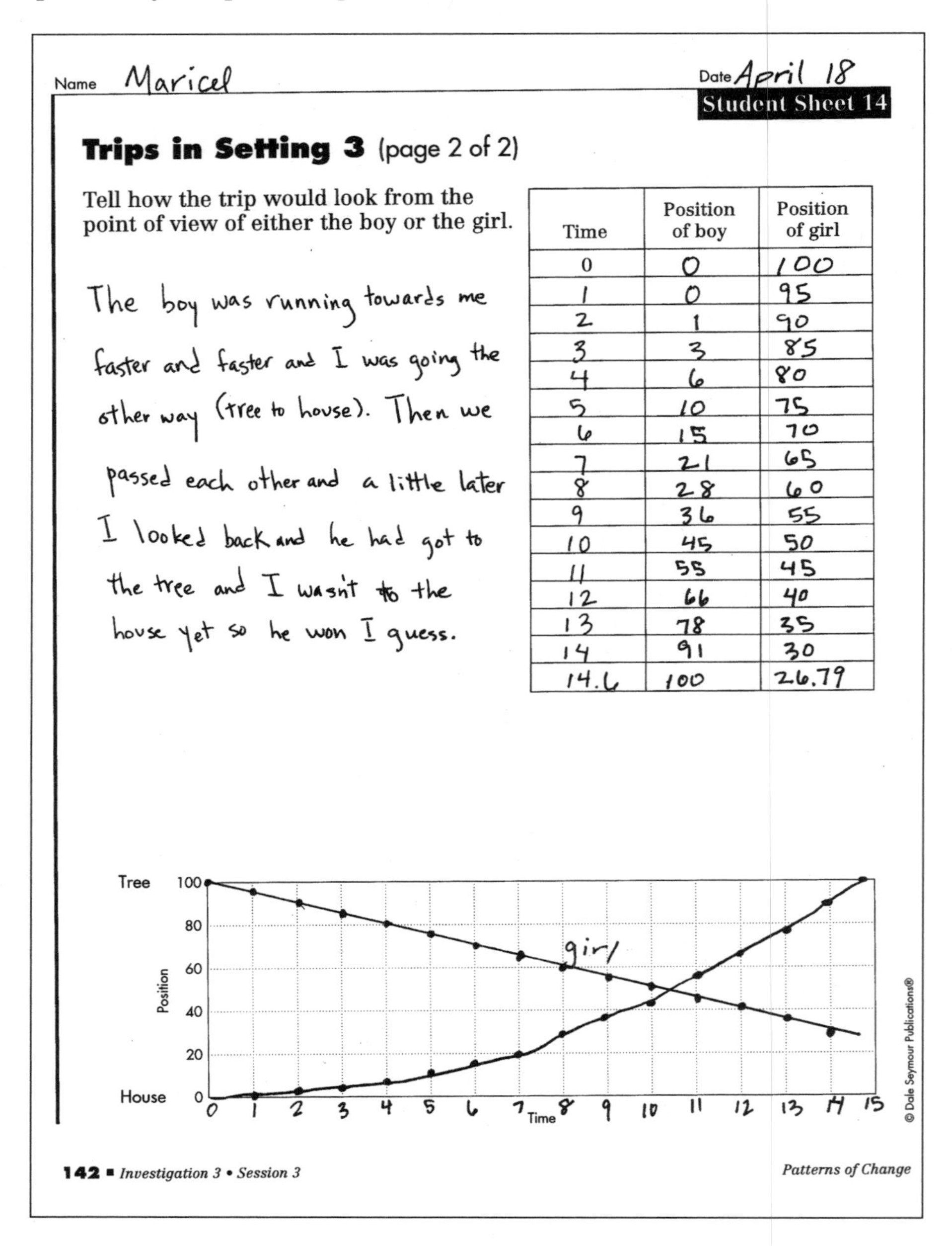

Name Maricel   Date April 18

Student Sheet 14

Trips in Setting 3 (page 2 of 2)

Tell how the trip would look from the point of view of either the boy or the girl.

The boy was running towards me faster and faster and I was going the other way (tree to house). Then we passed each other and a little later I looked back and he had got to the tree and I wasn't to the house yet so he won I guess.

Time	Position of boy	Position of girl
0	0	100
1	0	95
2	1	90
3	3	85
4	6	80
5	10	75
6	15	70
7	21	65
8	28	60
9	36	55
10	45	50
11	55	45
12	66	40
13	78	35
14	91	30
14.6	100	26.79

142 ■ *Investigation 3 • Session 3*   *Patterns of Change*

Session 3 Follow-Up

Hand out another copy of Student Sheet 11, Story of a Trip. Earlier, students created a trip for the boy and girl using constant step sizes. This time, they are to create a trip that involves a *change* of step size. As before, they write a description (story) of the trip, show their trip on the tracks, fill in the tables, and draw a position vs. time graph of the trip.

The Challenge The *Trips* software allows users to hide one or more of the windows on the screen: the main Trips window, the Table window, the Graph window, and the Command window. This enables students to find out how well they understand the relationships between the various representations of a trip.

For this activity, the class continues to work in groups. To start, half the groups create computer challenges, selecting which window they want to hide. The Command window must be hidden for all challenges. On a sheet of paper, students write down the setting they are using and the values of the starting positions, step sizes, and any changes in step size for the boy and girl.

To hide any window, students may either click on the close box in the upper left-hand corner of the window or select **Hide (name of window)** from the **Windows** menu. To reopen a window, they select **Show (name of window)** from the **Windows** menu. Another way to hide the girl and the boy is to select **Screen** on the **Options** menu, as described in the **Teacher Note,** About the *Trips* Software (p. 78).

Once the first challenges have been created, the other groups rotate through the computers, trying to list the commands and describe the hidden window to the challengers. Then the groups switch roles. At the end of the activity, students discuss all the challenges.

Which challenges were difficult? Which were easy? Why? What clues helped you reconstruct the missing windows?

Teacher Note — *Changing the Step Size in Setting 2*

When the students are working at the computer in *Trips* Setting 2, they may sometimes get decimal numbers in their table. This is probably because one step size was set to change at a point where the boy or girl figure does not land—a point that is stepped over.

Here's an example: Suppose the girl has a step size of 2 and is set to change her step size to 3 at position 5. With a step size of 2, starting from 0, she would never land on 5, stepping instead on 0, 2, 4, 6, and so on. If she has to change step size at 5, she cannot continue from 4 to 6, or go from 4 to 7.

Instead, at 4, the girl will take a half-step of the initial size. Half of 2 is 1, which will take her from 4 to 5. Her next half-step will be half of the new step size. Half of 3 is 1.5, which takes her from 5 to 6.5.

Thus, when her step size changes, the girl will go from 4 to 6.5, or a step size of 2.5, for just one step. Thereafter, her step size will be 3.

This may be confusing to students running trips at the computer, and those working at the metersticks may not know when to actually make the change of step size. The best way to avoid this problem is to suggest that students always change the step size at a position where the boy or girl will land. Alternatively, you might discuss the issue with the class so they have some sense of why decimal numbers sometimes appear in the table on the computer.

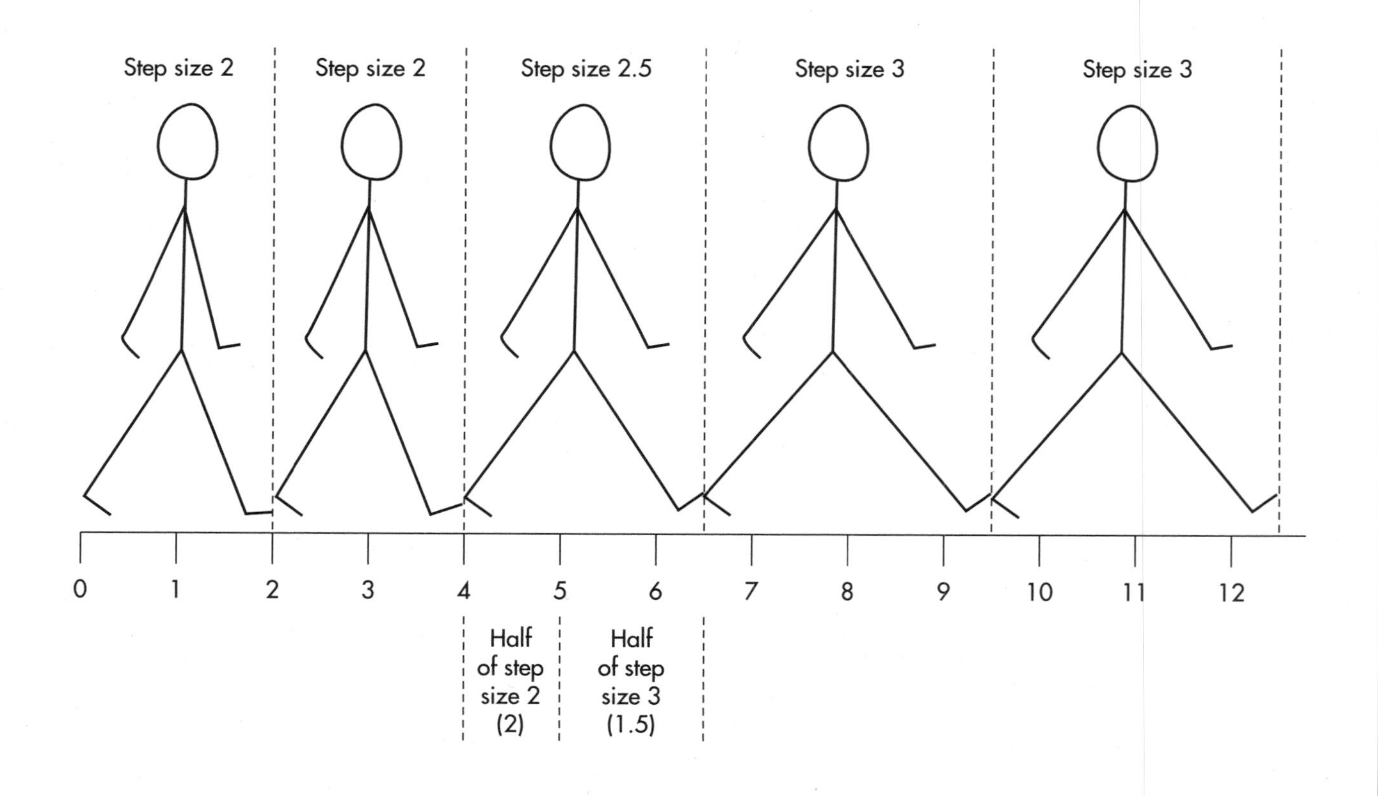

More Match-Ups

What Happens

Students match each other's tables and graphs (from Sessions 2 and 3) to the three motion stories, then as a whole group discuss any questions and observations about the match-ups. Student work focuses on:

- describing relationships between number tables, graphs, and motion stories

Materials

- Posted display of Three Motion Stories
- Students' completed stories, tables, and graphs from Student Sheets 10, 13, and 14
- Stick-on notes (available)
- Student Sheet 11 (1 per student, Extension)

Activity

Comparing Tables and Graphs

Collect the work students have done during this investigation on the second page of Student Sheets 10, 13, and 14. Then redistribute the sheets to pairs or groups of three, being sure that no one gets his or her own work back.

Allow time for groups to discuss the work and decide which of the three motion stories each sheet represents. Students post the sheets under the corresponding motion stories on display. Then give students time to look over the postings and add stick-on notes with any questions or comments they may have about the classification or about particular graphs, tables, or stories. Finally, gather the class together for discussion.

Are there any tables and graphs that don't seem to go with the story they are posted under?

Pick out particular examples from student work for discussion:

- **How do you know that these two tables both correspond to the same story?**
- **How do you know that these two graphs both correspond to the same story?**
- **How can you tell from this table what story it corresponds to? If the table and graph weren't on the same page, how could you tell that they go together? What features would you look for?**
- **How can you tell from this graph what story it corresponds to?**

Pick some tables or graphs that have stick-on notes on them, and discuss students' questions.

The aim of this discussion is to focus on ways of seeing the correspondences among motion stories, number tables, and graphs. It is important to realize that a single motion story can be expressed with infinitely many number tables and graphs; all of those tables and graphs, however, have to share some essential characteristics of the motion story. For example, for motion story 2, there must be a row in the number table under which the numbers indicating the girl's positions become greater than the numbers indicating the boy's positions, and the graphs of position vs. time have to cross.

Session 4 Follow-Up

Extension

Creating Challenges Distribute one more copy of Student Sheet 11, Story of a Trip. This time each student or pair makes up a story and marks the tracks for the story, leaving the tables and graphs blank. They turn in their work as a challenge for another student or pair, who must fill in the tables and draw a graph to go with the story.

The preparation of these challenges could be homework. Students would then exchange challenges in class, fill in the missing components, and show their solution to the challenger. The challenger assesses whether the proposed solution is correct, and why or why not.

Two Types of Graphs

What Happens

Students explore a different kind of graph, step size vs. time, discussing how to interpret different examples. Working in pairs, they prepare a set of distance vs. time and step size vs. time graphs to represent a "mystery" trip for the rest of the class to guess. Student work focuses on:

- making step size vs. time graphs
- exploring the relationship between position vs. time and step size vs. time graphs
- interpreting graphs of step size vs. time and position vs. time

Materials

- Student Sheet 15 (1 per student)
- Mystery Walk strips cut from Student Sheet 16 (1 strip per pair)
- Student Sheet 16 (whole copy, 1 per pair)
- Student Sheet 17 (1 per pair; optional extras for homework)
- Student Sheet 18 (1 per pair; optional extras for homework)
- Student Sheet 19 (1 per student)
- Scissors
- Glue or paste

Activity

Showing a Walk with Two Graphs

Note: When running trips on the computer, students deal with numbers for position and step size that are not specifically identified by a particular unit of measure; for example, a step size of 2, and a starting position of 15. When they use the meterstick and rods for a trip, the numbers correspond to centimeters. In Sessions 5 and 6, we link the numbers to meters. In order to show somewhat reasonable step sizes, the basic unit on each graph's vertical axis is 0.5 meter instead of 1 meter. Provide guidance as needed to clarify the decimal numbers. Also, in their graphing, students need to recognize that they will go up 2 "boxes" for 1 meter while they go right only 1 "box" for 1 second.

Distribute Student Sheet 15, Two Kinds of Graphs. Ask a volunteer to "walk" the first graph at one step per second, while the rest of the class gives suggestions and comments about the relationship between the graph and the walk.

It may be useful to discuss the corresponding table, raising issues such as the fact that position numbers can be large when the corresponding step size numbers are small—even 0.

Point out to students that on the second graph on Student Sheet 15, the vertical axis is labeled *step size* instead of *position.*

How big was the first step? (1 meter) **the second step?** (also 1 meter) **How big was the step taken during the third second?** (0 meters)

How can we see the size of each step on the graph of position vs. time? If we wanted to show the size of each step more directly, we could graph step size vs. time.

Suggest a bar graph in which the height of the bar for each second shows the size of the corresponding step. Discuss with students how this convention can be used to indicate step size. Then ask them to add bars to the step size graph to fit the story shown on the position graph.

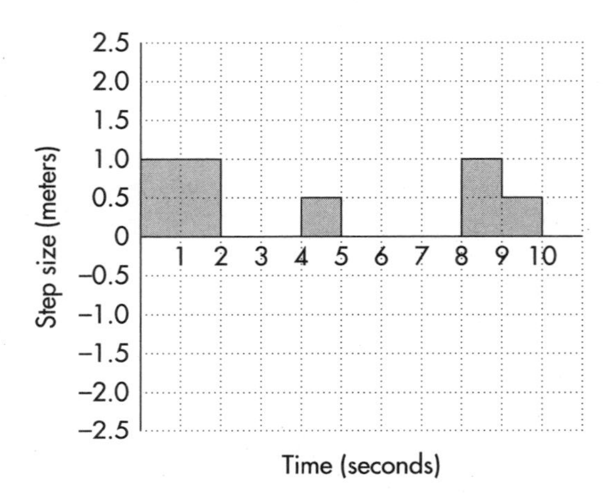

Ask students which of these graphs is more useful: step size vs. time or position vs. time. If they need a more specific question, ask:

Which of these graphs would be more useful to tell someone how to take a trip? Which graph would be more useful to determine how far you would go for a certain trip? Which graph would best help you to tell how fast you would have to go for a trip?

Can you reproduce a walk using either one of the graphs alone? Is it easy to create one graph from the other?

Follow the same procedure for the second page of Student Sheet 15, which describes a trip that includes a change of direction. Ask a volunteer to demonstrate the walk based on the position vs. time graph.

In the table, be sure students understand that step-size numbers are positive whenever the position numbers increase and negative when the position numbers decrease. (They will be familiar with this convention from the *Trips* program, where negative step-size numbers made the figures turn around and walk back.)

Students fill in the step size on the table and put bars on the second graph to show step size.

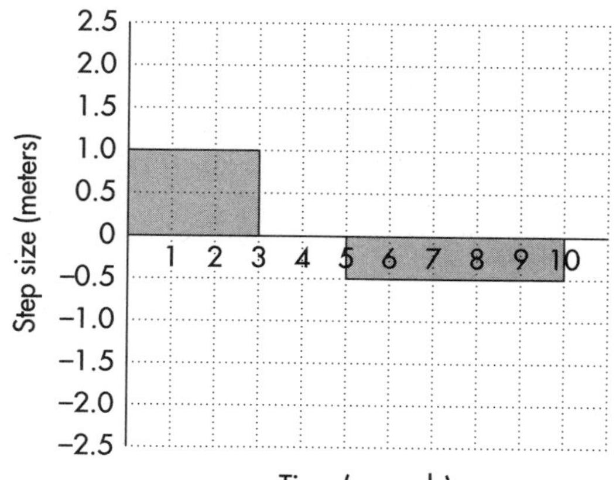

Note: The relationship between a position vs. time graph and its corresponding step size vs. time graph involves some of the most important ideas of the mathematics of change. As you discuss these graphs and the ideas they illustrate, it's important to stress the following dynamic aspects of that relationship:

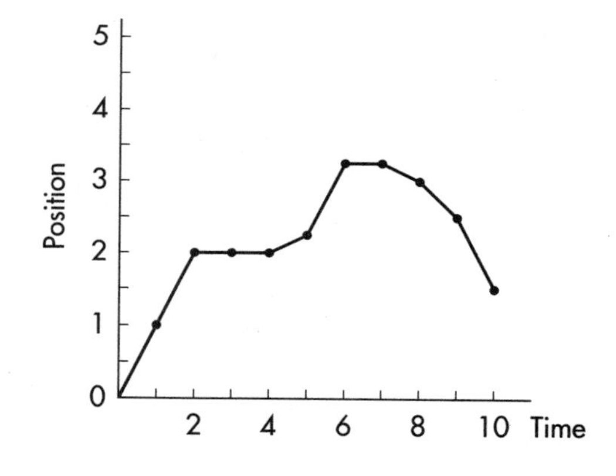

The steepness of the line in a position vs. time graph corresponds to height of a bar in a step size vs. time graph. The steeper the line in the position graph, the taller the corresponding bar or bars in the step-size graph.

The correspondence can be seen in terms of movement: the "up and down" movement of the bars on a step-size graph corresponds to the changing direction and steepness of the position line.

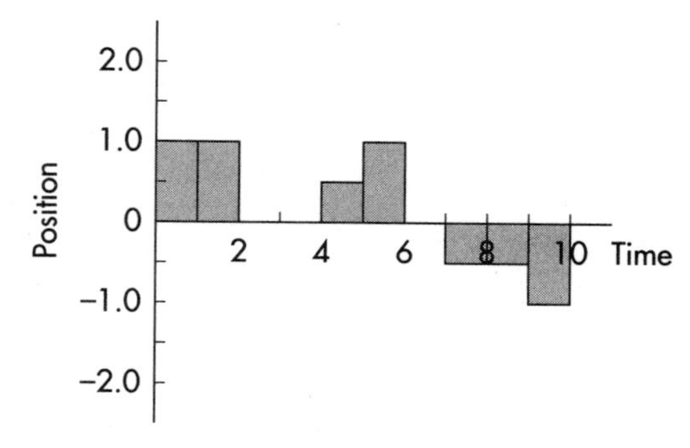

A position vs. time graph "going up" corresponds to a positive step size, and a position vs. time graph "going down" corresponds to a negative step size. A horizontal position vs. time graph corresponds to a zero step size.

Activity

Graphing and Guessing Mystery Walks

In this activity, students work in pairs making two kinds of graphs to go with one of the Mystery Walks. They will post their two graphs together, and the rest of the class will guess which Mystery Walk the graphs show.

Give each pair one of the Mystery Walk strips you have prepared. Students should be careful not to let anyone else to know which story they have. For their graphs, make available copies of the two blank graph templates: Student Sheet 17, Position vs. Time Graph, and Student Sheet 18, Step Size vs. Time Graph. Some students might create stories that don't fit on these graph templates; it's fine if they make their own.

Note: Some of the Mystery Walks include larger-than-normal step sizes; it is also likely that as students work out their graphs, they may include step sizes that are unrealistic for people. Keep in mind that the purpose here is to explore more generally the relationships between the two types of graphs; grant students creative license to include wide variation in step size, which will yield a greater variety of graphs.

After drawing their two graphs, student pairs post them on the wall, one just below the other. When their graphs are posted, they come to you to get a copy of *all* the Mystery Walks (Student Sheet 16). Allow time for everyone to look at the posted graphs and figure out which Mystery Walk each pair of graphs matches.

Then gather the class together. Student pairs take turns showing the graphs they have drawn while their classmates discuss which is the corresponding story. Any student who makes a guess must explain his or her choice. Encourage students to discuss several possibilities before the pair that is presenting their graphs reveals their Mystery Walk. Ask the presenters and the class the following questions:

Are these graphs a good match for their Mystery Walk story? Are any other stories that have been suggested also a possible match for these graphs?

Activity

Assessment

Graphs: What Story Do They Tell?

Give each student a copy of Student Sheet 19, What's the Story? Have available scissors and glue, paste, or tape.

On the first page here, you'll see some graphs that show walking trips. There are three graphs showing how position changes throughout the walk, and four graphs showing how step size changes.

Your first task is to match each position graph to the step-size graph for the same trip. Cut out the graphs and put them in matching pairs. There will be one extra step-size graph that doesn't go with any of the position graphs.

Circulate and observe students as they work. Some features that students should notice in pairing graphs are the following:

- The steeper the line on the position graph, the higher the bar on the step-size graph, and the faster the motion. Similarly, less steeply sloped lines on the position graph correspond to shorter bars on the step-size graph, and slower motions.
- A horizontal line (the least steep line possible) on the position graph corresponds to a 0-height bar (no bar at all) on the step-size graph, and to a stop in the motion.
- When the line on a position graph goes up, the step size bars are positive (above the horizontal axis), and the motion is forward. When the line on the position graph goes down, the step-size bars are negative (below the horizontal axis), and the motion is backward.

Allow 5 to 10 minutes for this task before continuing. Then direct attention to the second page, and ask for volunteers to read the three stories aloud.

Decide which pair of graphs from the first page goes with each motion story. Paste the graphs under that story. Then, beside the graphs, explain in writing how you would convince someone why these two graphs go with this story. Describe the specific features of each graph that tell you that it goes with the story.

❖ **Tip for the Linguistically Diverse Classroom** Students who are not writing proficiently in English might make their explanation orally, pointing to specific parts of the graph to support their thinking.

Advise students that you are looking for more explanation than "I guessed that this was the right story" or "There was only one pair of graphs left, so I picked them." If there is enough time, students can work on this in class. Otherwise, they can do the writing for homework.

If some students finish early, ask them to imagine that someone disagrees with their choices. Suggest that they also write the reasons why none of the graphs except the pair they picked could go with each story.

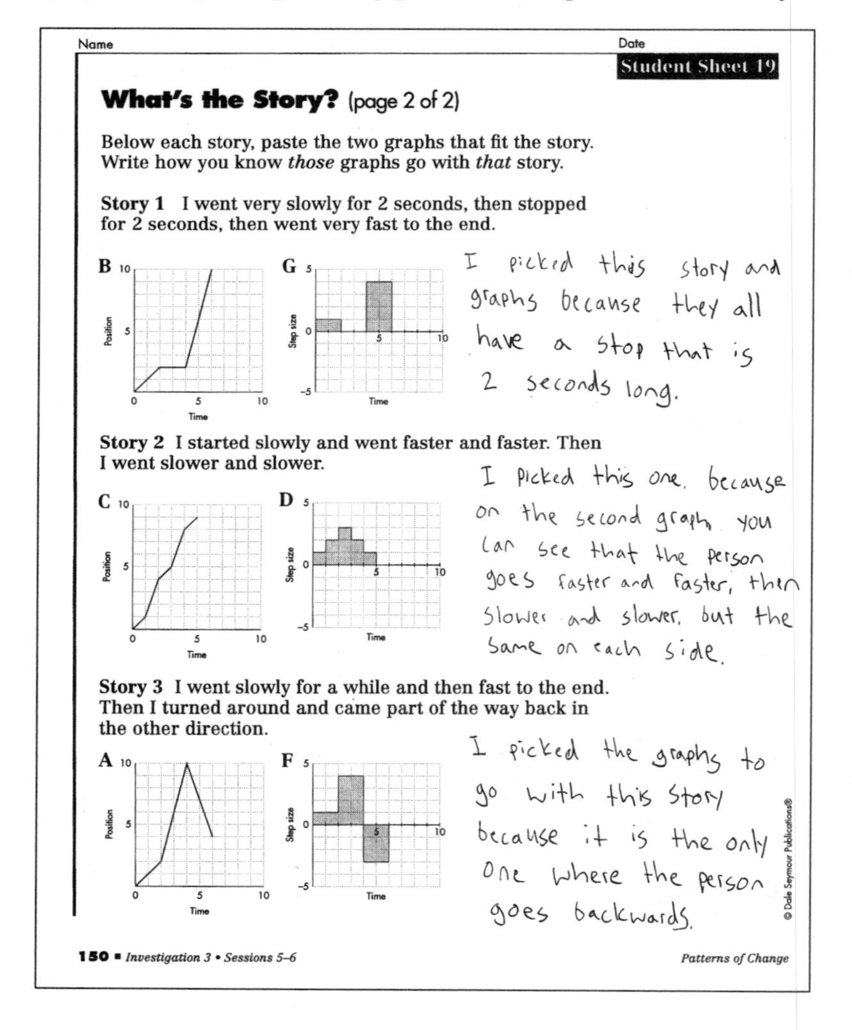
Name Date

Student Sheet 19

What's the Story? (page 2 of 2)

Below each story, paste the two graphs that fit the story.
Write how you know *those* graphs go with *that* story.

Story 1 I went very slowly for 2 seconds, then stopped for 2 seconds, then went very fast to the end.

I picked this story and graphs because they all have a stop that is 2 seconds long.

Story 2 I started slowly and went faster and faster. Then I went slower and slower.

I picked this one. because on the second graph you can see that the person goes faster and faster, then slower and slower, but the same on each side.

Story 3 I went slowly for a while and then fast to the end. Then I turned around and came part of the way back in the other direction.

I picked the graphs to go with this story because it is the only one where the person goes backwards.

150 ▪ *Investigation 3 • Sessions 5–6* *Patterns of Change*

In their writing, encourage students to describe in some detail how they related features of the graphs to parts of the story. For example:

> In this story, the person walked slow, stopped, and then walked faster. So, I chose graph B that has a flat place for a stop. It has the second slanted line steeper than the first slanted line because the person went faster then. I also picked Graph G that had small steps, then no steps for a stop, then bigger steps for faster. No other graph has a stop, except for the one that goes backward.

Sessions 5 and 6 Follow-Up

Students draw graphs of position vs. time and step size vs. time for the following motion story:

> The lioness was hunting. She walked very slowly for 4 seconds. Then she stood perfectly still for 5 seconds. Suddenly, she pounced, moving very fast for 3 seconds.

They may use their own paper, or you might provide extra copies of Student Sheets 17 and 18.

Comparing Graphs on the Computer Screen Students use the *Trips* program to run some trips on the computer with graphs of step-size vs. time instead of position vs. time. This is done by choosing **Graph Step** from the **Options** menu.

Suggest that students first plan a trip and sketch a step-size graph for that trip. After they run the trip, they can compare the computer's graph with their sketch.

Session 7 (Excursion)

Animation

Materials

- Demonstration flip book
- Stick-on notes in light colors (1 pad per student)

What Happens

Students create two animated flip books, one showing a change over time, and another showing two simultaneous changes that take place at different rates. Students discuss the relationship between successive differences and rate of change. Their work focuses on:

- showing relative change or motion

Activity

Animated Flip Books

Show students the flip book you have made and ask what change it demonstrates. Brainstorm with the class other kinds of changes over time that they could show in a flip book that they will be drawing. Encourage them to imagine motions that would be simple to draw, such as a ball bouncing, a stick falling, or a wheel turning.

The students will work individually to produce a flip book that shows some kind of change over time. Suggest that they design books of 10 to 20 pages, tearing off this many notes (as a chunk) from their pad of stick-on notes. Explain that it is helpful to start from the back of the flip book and work forward; that way they can see through each page the image that follows. Stress the importance of using simple drawings so students can create their books in reasonable time and without unnecessary complications.

After they finish their flip books, they exchange them and identify the change over time that each book shows.

Successive images from a student's flip book

Gather the class to share their experiences.

What was difficult about making the flip books? What was easy? Did your readers figure out the change that you had in mind? How do you make something that changes quickly or slowly?

Activity

Flip Books with Two Changes

This activity is more challenging: Students create a flip book that shows two things changing together, but at different rates.

Note: In a flip book, the rate of change is affected by two factors: the flipping rate (the more pages flipped per second, the faster the change seems to occur), and the differences between successive images (the more similar they are, the more slowly the change seems to occur). Since the flipping rate will be the same for both of the elements in these flip books, students should focus on the differences between successive images.

Write three possibilities on the board:

> Story A: One thing changes slowly, and another thing changes fast.
>
> Story B: One thing changes at a constant rate, and another thing starts to change slowly and then gets faster.
>
> Story C: One thing changes at a constant rate, and another thing changes quickly at first and then slowly.

Each student chooses one of these three possibilities and designs a flip book that shows the story as clearly as possible. After students complete their flip books, they exchange them with a partner. Each student tries to recognize the story portrayed in the flip book they received—not only which type of story is shown (A, B, or C), but also the specific changes. For example, "It's story A. The sunflower grows tall really fast while the boy grows hardly at all."

You might collect the flip books for an exhibit that students and classroom visitors can explore at another time.

Activity

Choosing Student Work to Save

As the unit ends, you may want to use one of the following options for creating a record of students' work on this unit.

- Students look back through their folders or notebooks and write about what they learned in this unit, what they remember most, and what was hard or easy for them. Students might do this during their writing time.
- Students select one or two pieces of their work as their best, and you also choose one or two pieces, to be saved in a portfolio for the year. Students can create a separate page with brief comments describing each piece of work.
- You may want to send a selection of work home for families to see. Students write a cover letter, describing their work in this unit. This work should be returned if you are keeping year-long portfolios.

Session 7 Follow-Up

Homework

Students may choose one of the following possibilities:

1. Show your second flip book to someone in your house. Ask that person to guess the story and talk about the differences in how fast the changes take place. Write about your experience.
2. Make up another motion story, and make a flip book to go with it.

Nearest Answer

Basic Activity

Students estimate answers to computation problems by rounding numbers in the problems and computing mentally. They pick the closest answer among the choices that are provided. In a variation, students choose an approximate number for a place marked on a number line between two given numbered points.

Nearest Answer provides practice with rounding numbers and estimating answers. This kind of thinking helps in checking answers found by calculator. Students' work focuses on:

- rounding numbers
- calculating mentally
- comparing possible answers to find the closest one

Materials

- Overhead projector
- Overhead transparencies of the problems you will use in the session; choose from the examples in the discussion that follows, or design similar problems yourself.
- Pieces of paper or cardboard for covering parts of the problems
- Calculators (optional)

Procedure

Step 1. Prepare a problem and four answer choices. Keep it hidden from the students. If you are writing your own, include as one answer a fairly round number that is a good estimate, and three other answers that you think might be tempting if students are not thinking carefully. For example:

2,897,897 + 37 =
5,000,000 3,000,000 2,000,000 29,000,000

Tell students that you are going to show them an arithmetic problem for only a few seconds. They are to round the numbers in the problem to make them easier to compute with, and estimate the answer.

Note: If an overhead is not available, problems can be written on the board or chart paper.

Step 2. Present the problem, keeping the answers covered, for 20 to 30 seconds. Decrease this time to 15 seconds as students become accustomed to the activity and problem type. It is important not to show the problem so long that students have time to work it out in writing.

Step 3. Cover the problem, and show the choice of answers. Students write down the answer they think is closest. (In the example problem, they might round the numbers to 3,000,000 + 0, or 2,900,000 + 40, and choose 3,000,000 as the closest answer.)

Step 4. Uncover the problem and discuss. One or two students tell how they rounded and why they chose their answer.

Following are some whole-number problems to get you started. Plan to supplement these with problems that you or your students write.

29 + 52 =
40 60 80 100

---

545 – 240 =
200 300 400 700

50,102 – 2898 =
10,000 20,000 40,000 50,000

32,010 – 934 =
12,000 23,000 31,000 51,000

36,010 – 19,999 =
1600 16,000 18,000 56,000

5210 + 298 =
5400 5500 7000 8000

591,000 + 211,000 =
700,000 800,000 900,000 10,000,000

3,928,012 – 43 =
28,000 350,000 3,000,000 4,000,000

3,051,860 + 815 =
5,000,000 4,000,000 3,000,000 2,000,000

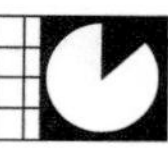

$7108 - 141 =$			
5000	6000	7000	8000
$5982 + 978 =$			
6000	7000	14,000	15,000
$608 \times 980 =$			
5000	50,000	600,000	690,000
$9 \times 211 =$			
20	200	2000	20,000
$50,300 \div 4926 =$			
1	10	100	1000
$59 \times 11 =$			
60	500	600	6000

Variations

Nearest Answer Decimal Problems Students round the decimals to the nearest whole number or, for large numbers, to a landmark number. For example, 527.9 – 2.1321 can be thought of as 528 – 2 or 530 – 0. The accuracy needed will depend on the answer choices. Sample problems:

$1.1 \times 54 =$	5.4	54	540	5400
$342 + 0.999 =$	14,000	13,000	12,000	340
$82 \div 4.2 =$	0.5	2	20	40
$24.8 + 3.1 =$	29	28	27	270
$498 \times 10.13 =$	5.00	50.0	500	5000
$59.3 \times 1.1 =$	60	600	6000	60,000
$435.4 \div 0.98 =$	4.4	44	440	4400
$268 \div 9.9 =$	25	250	2.5	2500
$402 \times 2.96 =$	400	800	1200	8000
$25 - 2.1 =$	4	12	23	30
$4.3 - 1.412 =$	0	1	3	6
$29.93 - 2.1 =$	9	20	25	28
$80.5 \div 3.97 =$	4	10	20	80
$311 + 3.71 =$	11	300	600	800

Nearest Answer Fraction and Mixed Number Problems Students round the fractions to the nearest whole number or, occasionally, to one-half, and estimate. Sample problems:

$8\frac{1}{13} \times 2\frac{9}{11} =$	16	18	24	64
$15\frac{7}{8} + 2\frac{6}{7} =$	17	18	19	29
$5\frac{9}{11} - 2\frac{7}{8} =$	2	3	$3\frac{2}{3}$	32.3
$3\frac{7}{8} + \frac{1}{15} =$	3	4	38.23	50
$\frac{32}{66}$ of 22 =	$\frac{1}{3}$	7	10	45
$\frac{3}{4}$ of 83 =	20	60	240	560
$\frac{3895}{39} =$	0.10	10	100	1000
$\frac{1}{11} + \frac{8}{9} =$	1	2	10	18
$\frac{1}{3} + \frac{4}{7} =$	$\frac{1}{2}$	1	2	3
$\frac{11}{4} =$	0.5	2.8	15	44

Nearest Answer Percent Problems Students use a nearby familiar percent to help them choose an answer. For example, 26% of 77 could be thought of as close to 25% (or 1/4) of 80. Sample problems:

33% of 15.85 =	5	45	450	500
198% of 15 =	7.5	15	30	3000
51% of 69 =	16	35	4	100
98% of 14.3 =	1400	143	14	0.143
25.9% of 774 =	0.2	2	20	200
73% of 406.2 =	50	100	200	300
24% of 83.6 =	2	20	27	100

A bicycle listed at $210 is on sale at a 30% discount. The sale price is about:

$70	$150	$180	$200

With an increase of 10% per year, an article now costing $49 may be expected, in 12 months time, to cost:

$54	$57	$100	$200

Nearest Answer Number Line Problems A number line is provided with three points labeled—two with numbers, and the third with the letter A. Students decide what number A is nearest to. Sample problems:

Number line	A is nearest:			
2 A 5	3	3.5	4	4.5
0.7 A 0.8	0.6	0.15	0.76	0.79
0 A 1	−1	1/4	2/5	3/4
−3 −1 A	−5	−2	0	1
6 A 10	6	7	8	9
8 A 9	8.5	8.21	8.36	8.7
800 A 900	8.6	8.7	840	870
580 A 590	5.82	58.2	582	

Comparing Estimation in Addition and Multiplication Pose addition and multiplication problems that use the same numbers. Working in pairs, students decide what four or five answers to provide for other students to choose among. They are likely to find that the answers for addition problems cannot range in size as much as answers for multiplication problems if they are still to stump people. For example:

$726 + 977 =$ 1700 1800 1900 8100

$726 \times 977 =$ 8400 80,000 700,000 800,000

Looking at the Effects of Rounding Different Factors Pose multiplication problems with the requirement that students round only one of the factors to a number they can multiply by mentally. Students investigate this using calculators. Which number should they round? By how much can they round and still get a reasonably accurate answer?

Students will probably find that if the factors are close in size, rounding either factor has approximately the same effect. For example, for 38×52 [answer 1976], either 38×50 or 40×52 will give a reasonable approximation. Rounding both numbers to 40×50 gives a closer approximation because the numbers are rounded in opposite directions.

However, if the factors are of very different size, rounding the larger factor will result in a closer answer than rounding the smaller number. For example, for 8×389 [answer 3112], 8×400 produces a closer answer than 10×389, even though only 2 was added to the 8, whereas 11 was added to the 389.

Related Homework Options

- **Students Invent Their Own Problems** At home, students prepare problems with a choice of answers. They write about how they would do their own problems. Guide students to use numbers that are near landmark numbers or, in the case of fractions, near whole numbers. At another Ten-Minute Math time, students might exchange their problems with others and compare strategies.
- **Page of Problems to Do at Home** Students take home two copies of a page of six to eight problems with answer choices. They do these with a family member or friend as quickly as they can, and then take all the time they need to check their answers, in writing or with a calculator. This provides an opportunity for students to explore a new kind of problem in depth.

Graph Stories

Basic Activity

Students look at a graph shape and try to imagine the story it is telling. In looking at graph shapes without numbers on either axis, students interpret qualities of the shape: relative heights, steepness of slope, pointed or gradual turns, going up or going down. Student work focuses on:

- attending to important features of a graph
- imagining the stories behind graphs that show change over time
- drawing a graph to fit a particular story

Materials

- Overhead projector
- Graph shapes on transparency film
- Squares of transparency film (for students to draw their own graph shapes)

Procedure

Step 1. Show a graph on the overhead. Choose one of those provided (p. 152) or another that you or a student has made. Ask questions to start students thinking: "What could be happening in this graph? What could be changing as time goes by? What might be growing and shrinking? going faster or slower? becoming more or less?"

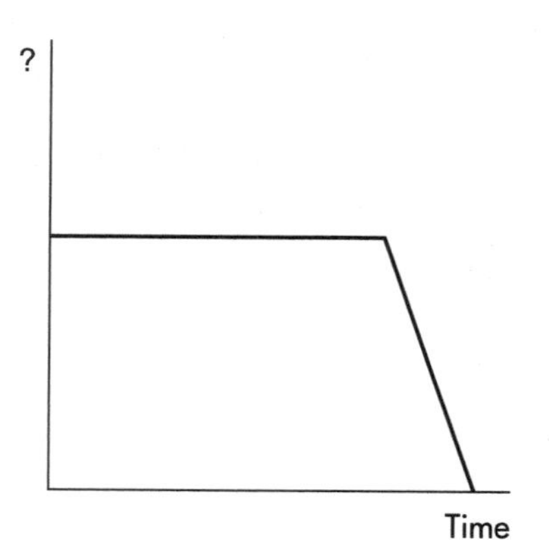

Step 2. Students talk in pairs about stories that might fit the graph. They could each invent one story, being sure it makes sense to the other person, or partners might collaborate on one story. In their story, students tell what variable is being shown and how it changes.

Step 3. Share students' stories. A few students tell their stories, explaining or showing how the story fits the shape. There are endless possibilities. For example:

- The graph shows how fast I went: I started running very fast, but then I got tired and slowed down until I stopped.
- The graph shows how much sunlight there was: There was a lot of sunshine in the room. Then later in the afternoon a storm came and the room got darker and darker until it was night.
- The graph shows how many people: A lot of people were in the room. People started leaving. People kept leaving until there was no one left.

Variations

Making Graphs to Go with Stories Prepare (or ask a student to prepare) a story; have in mind a graph that fits it. Tell the story, or show it on the overhead for students to read. Students draw graph shapes that fit the story, perhaps on pieces of transparency so they can be easily shared. Share the graphs in small groups and then a few of them with the whole class.

Making Graphs to Describe a Personal Change On a topic you select, students draw a graph of their changing feelings. Topics might include these:

- How awake I feel as I go through a day.
- How much I have liked school since the beginning of the year.
- How hungry I feel as I go through the day.
- How I have felt about math over my years at school.
- How my skill at (playing an instrument, playing a sport, reading, drawing, or the like) has changed over the years.

Comparing Stories for Two Graphs Draw two graphs on the same grid. Students tell stories about the graphs that bring out the similarities and differences. For example:

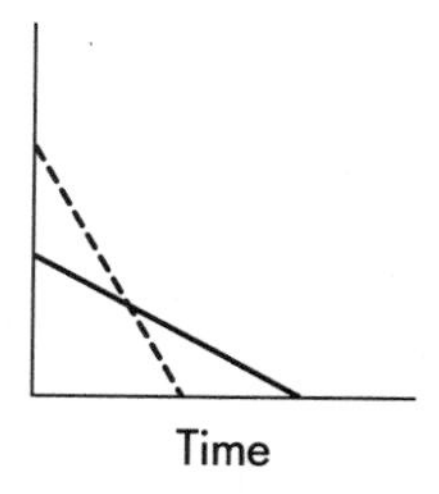

This is the food in our dogs' dishes at mealtime. The dotted line is the food in Tipsy's dish. She gets more food and eats it superfast. The plain line is the food in Max's dish. He eats slowly and we have to keep Tipsy from pushing him away and getting her nose in his dish.

Related Homework Options

Making Up Stories Students take home a graph shape and write one or more stories to go with it. They might enlist family members' help in thinking up possible stories.

Inventing Graphs and Stories That Go Together Students write brief stories of something changing and make a graph to go with it. You might use some of these stories for the variation Making Graphs to Go with Stories.

The following activities will help ensure that this unit is comprehensible to students who are acquiring English as a second language. The suggested approach is based on *The Natural Approach: Language Acquisition in the Classroom* by Stephen D. Krashen and Tracy D. Terrell (Alemany Press, 1983). The intent is for second-language learners to acquire new vocabulary in an active, meaningful context.

Note that *acquiring* a word is different from *learning* a word. Depending on their level of proficiency, students may be able to comprehend a word upon hearing it during an investigation, without being able to say it. Other students may be able to use the word orally, but not read or write it. The goal is to help students naturally acquire targeted vocabulary at their present level of proficiency.

We suggest using these activities just before the related investigations. The activities can also be led by English-proficient students.

Investigation 1

grow, shrink

1. Blow a balloon up a little way, commenting that it is growing. Do this a few more times until the balloon is inflated.
2. Ask students how to make the balloon shrink. Let the air out a little at a time and have students use the word *shrink* as they describe what is happening. Use a large rubber band, stretching it out and letting it shrink back to reinforce the idea if needed.
3. Ask if students can name other things that grow and shrink (clothes can shrink, children grow, the moon grows and shrinks alternately).

change, speed, faster, slower, increase, steady

1. At a time when you can take the students outside, invite them to play Follow the Leader on a trip with changing speeds. Model for them a moderate walking speed, a slower walk, and a faster walk, using the vocabulary above.
2. Next, model changes from moderate to slow to fast, calling out, for example, "Change speed; go slower," then, "No change; steady speed," then, "Change speed," or "Increase speed; go faster."
3. Put groups of students through their paces, asking them to walk according to the patterns you call out to them. Let students proficient in English help to model the paces and changes as you use the vocabulary.

steep, flat, slope

1. Draw on the board sketches of two hills, one with a steep slope and one with a gentle slope, and a horizontal line. Identify the hillsides as being steep or not steep and as having a steep slope or a slope that is not as steep. Identify the line as flat.

2. If there are steep and gentle hills in your area, discuss them with the students, asking, for example, which ones are easier to walk up, or what it is like to ride a bike down them. How do the flat stretches compare with the hills?

Investigation 2

position, start, end

1. Direct a group of about five students to form a line. Name the first student in line and say that he or she is at the *start* of the line. Name the last student and say that he or she is at the *end* of the line. Ask other students to repeat what position each of those students has.
2. When you call "Change places," the students change their order in the line. Discuss who now has the *start* position and who has the *end* position in the line. You might assign a number to each student in line and ask the class to name the student in position 1 (the start), in position 2, and so on, to whatever is the last position (the end).

How to Install *Trips* on Your Computer

The disk packaged with this *Patterns of Change* unit contains the *Trips* program. You can run the program directly from this disk, but it is better to put a copy of *Trips* on your hard disk and store the original disk for safekeeping. Putting a program on your hard disk is called *installing* it.

Trips runs on a Macintosh II or above, with 4 MB of internal memory (RAM) and Apple System Software 7.0 or later.

To install the contents of the *Trips* disk on your hard disk, follow these steps (or the instructions with your computer).

1. Lock the *Trips* program disk by sliding up the black tab on the back, so that the hole is open. The *Trips* disk is your master copy of the program. Locking the disk allows copying it while protecting its contents.

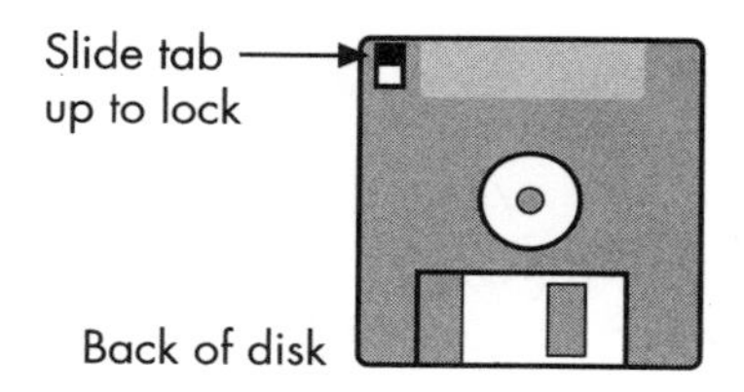

2. Insert the *Trips* disk into the floppy disk drive.
3. Double-click on the icon of the *Trips* disk to open it.
4. Double-click to open and review the Read Me file for any recent changes in how to install or use the program. Click in the Close box after reading.
5. Click on and drag the *Trips* disk icon (the outline moves) to the hard disk icon until the hard disk icon is highlighted, then release the mouse button.

   A message appears, indicating that the contents of the *Trips* disk are being copied to the hard disk. The copy will be in a folder on the hard disk with the name *Trips*.
6. Eject the *Trips* disk by selecting it (click on the icon) and choosing **Put Away** from the **File** menu. Store the disk in a safe place.
7. If the hard disk window is not open on the desktop, open it by double-clicking on the hard-disk icon. The hard disk window will appear, showing you the contents of your hard disk.

   Among its contents, you should see the folder labeled *Trips,* holding the contents of the *Trips* disk.
8. Double-click the *Trips* folder to select and open it. When you open this folder, the window shows the program your students will be using with this unit.

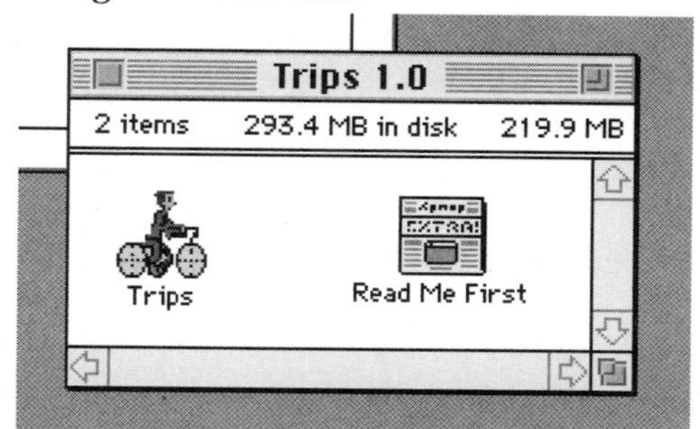

To select and run *Trips,* double-click on the program icon.

Optional Use of Alias

For ease at startup, you can create an alias for the *Trips* program by following these steps:

1. Select the program icon.
2. Choose **Make Alias** from the **File** menu. The alias is connected to the original file that it represents, so when you open an alias, you are actually opening the original file. This alias can be moved to any location on the desktop.
3. Move the *Trips* alias out of the window to the desktop space under the hard disk icon. For startup, double-click on the *Trips* alias. This is simply a shortcut that saves you from having to open first the hard disk and then the *Trips* folder to start the program inside.

Using the Menus in *Trips*

When you have opened *Trips,* the words you see across the top of the screen are called the *menu bar.* To see the items in a menu, point to the menu name with the on-screen arrow and press the mouse button. Holding the mouse button down, drag the selection bar down to choose different items on the menu. When your choice is highlighted, release the mouse button to select that item.

Some menu choices are also available from the keyboard. On the menu, the ⌘N indicates that, instead of selecting **New Work** from the menu, you could enter ⌘N. Hold down the command key (with the ⌘ and symbols on it), then press the N key.

A menu choice may sometimes be dimmed, indicating it is not available in a particular situation.

The File Menu

The **File** menu deals with documents and quitting.

File

New Work	⌘N
Open My Work...	⌘O
Close My Work	⌘W
Save My Work	⌘S
Save My Work As...	
Page Setup...	
Print...	⌘P
Quit	⌘Q

New Work starts a new document.

Open My Work... opens previously saved work.

Close My Work closes present work.

Save My Work saves the work.

Save My Work As... saves the work with a new name or to a different disk or folder.

Page Setup... allows you to set up how the printer will print your work.

Print... prints your work.

Quit gets you out of *Trips* altogether, if you want to do something else on the computer or if you are ready to shut down the computer.

Saving Your Work

At any time while you are working in *Trips,* you can save your work. When you save your work for the first time, a dialogue box opens. Enter a name in the box below the words **Save as.** To identify your work later, it is generally a good idea to include your name or initials, a brief description of the work, and the date; for example: `Doug C. Trip 1  3/24`

After the name is typed in, click on the Save button. For the remainder of that session, you can save your work simply by selecting **Save My Work** from the **File** menu or pressing ⌘S (Command-S), and the list of commands for your trip will be saved.

If you are sharing the computer with others and it's their turn, save your work, then choose **Close My Work.** Later, to resume your work, choose **Open My Work...** and select the work you saved.

When you quit after working in *Trips,* you will be asked if you want to save your work. If you have not saved already and you want to save, follow the same steps as above.

To save work on a floppy disk, see p. 116.

The Edit and Font Menus

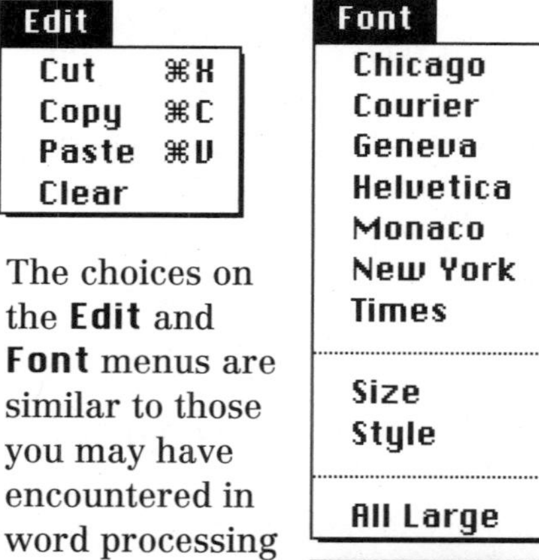

The choices on the **Edit** and **Font** menus are similar to those you may have encountered in word processing programs.

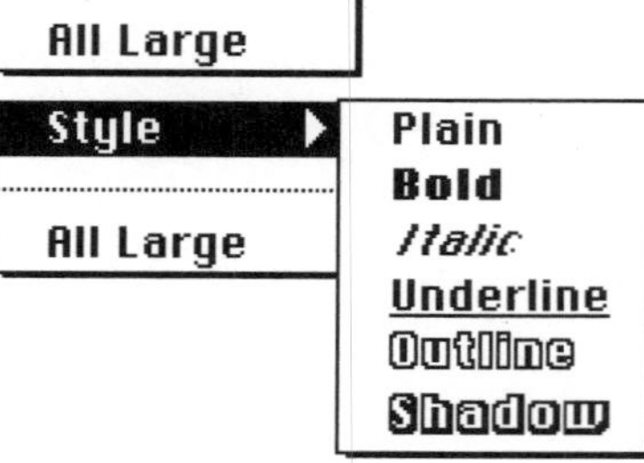

The **Edit** menu is used to change text: **Cut** deletes the selected text and puts it on the clipboard. **Copy** puts selected text on the clipboard. **Paste** puts the text on the clipboard into the active window. **Clear** deletes the selected text but does not put it on the clipboard.

If you are using the Notes window, you may want to use the **Font** and **Style** menu to change the typeface, size, and style of your text.

All Large changes the text in the Command, Notes, and Print windows to a large-size font. This is useful for demonstrations. This selection toggles (changes back and forth) between **All Large** and **All Small.**

The Windows Menu

Trips has more individual windows than many programs you may have worked with. The Trips window is the largest one; in it, the boy and girl take trips along tracks from the house to the tree. The words you see at the top of this large window depend on the setting you have chosen (Setting 1, 2, or 3); we call it the Trips window for simplicity. The Command window is where students work to set up different trips. The Graph window will graph a trip, and the Table window completes a table of the changing positions over time for both the boy and the girl.

The **Windows** menu allows you to show or hide any of these windows separately. The selection toggles between **Hide [Name of Window]** and **Show [Name of Window].** That is, when you select **Hide Trips,** the menu item changes to **Show Trips.** You can also hide a window by clicking in the "close box" in the upper left-hand corner of the window. To make it reappear, you need to use the Windows menu.

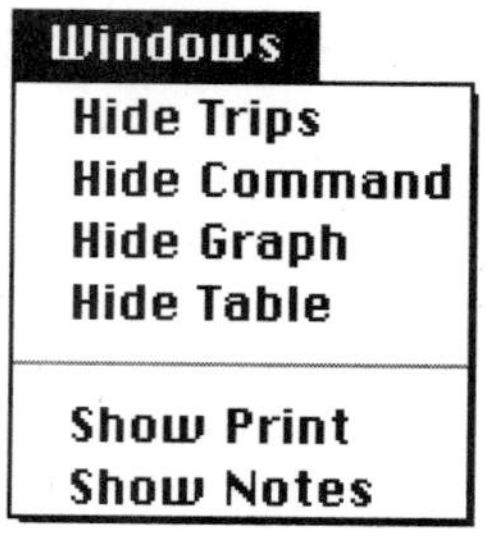

Windows

Hide Trips
Hide Command
Hide Graph
Hide Table

Show Print
Show Notes

The Print window is usually hidden. This is where the commands "print" and "pr" put their text.

The Notes window (initially hidden) is a word processor that can be opened and used while you are in *Trips*. Students might use it, for example, to describe their strategy for solving a certain problem, or to write a note about how they plan to continue the next day.

Settings and Options Menus

Each of the three settings in the **Settings** menu selects a set of commands that are placed in the Command window. The three settings are described in the **Teacher Note,** Using the Three Trips Settings (p. 85).

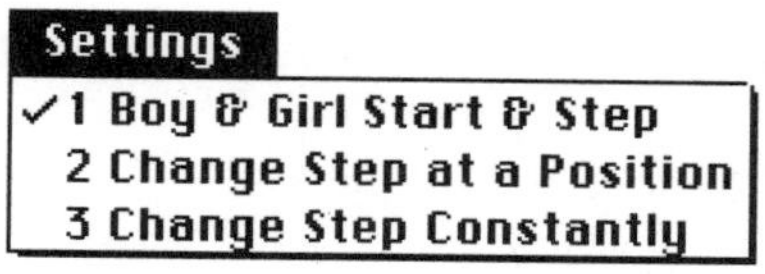

Settings

✓1 Boy & Girl Start & Step
2 Change Step at a Position
3 Change Step Constantly

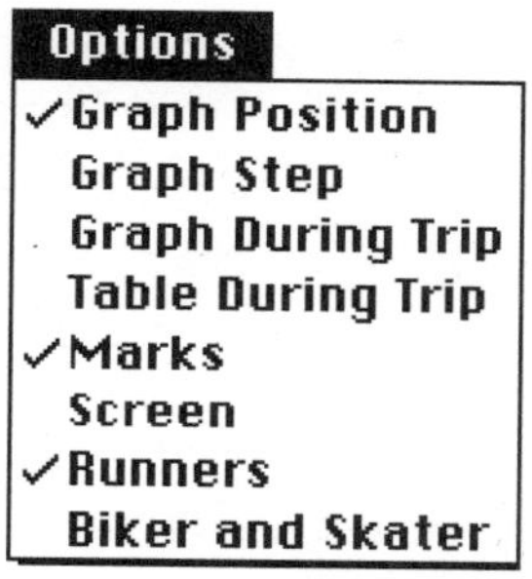

Options

✓Graph Position
Graph Step
Graph During Trip
Table During Trip
✓Marks
Screen
✓Runners
Biker and Skater

The **Options** menu allows you to customize *Trips*. For example, you may choose whether you will graph position or step size over time. You can also choose whether you will have the graph drawn and table filled in while the boy and girl are moving on their trip, or only after the trip is completed. A check mark indicates which options you have selected.

You also have the option of showing or not showing the arrows (**Marks**) that appear below each track at certain intervals throughout a trip. These are initially set to appear every 2 steps. You can change the interval between marks with a command; for example, `marksevery 4` gives you an arrow every four steps.

Screen offers the option to place screens over the tracks, hiding the boy and girl.

You can choose either **Runners** or **Biker and Skater** to change the animation used for the boy and girl.

The Help Menu

The Help menu provides some on-screen assistance while you are working in the *Trips* program.

Help
Windows...
Vocabulary...
Hints... ⌘H

Windows... shows an illustration of the screen with basic information explaining the Command window, the Graph window, and the Table window.

Vocabulary... provides a dictionary of the commands in *Trips,* also printed in this Computer Help section (see below).

Hints... gives a series of hints on the present activity, one at a time. It is dimmed when there are no available hints.

Making Your Own Trip: Commands

Whatever setting you are working in, you can change the commands in the Command window to change the trip. You can also add commands.

You need not start a new line for each command. However, when you want to start a new line, hitting the **<return>** key simply reruns the trip. After the trip, the cursor appears on a new line after your last command. If you want to open a new line in the middle of the existing commands, place the cursor at the end of the preceding line and press ⌘L.

Some possible commands are described below. While you are working in *Trips,* choose **Vocabulary** from the **Help** menu to get a listing of the commands you can use.

`always`

Used in a "*when* list" (see p. 115), this causes the action given to be taken at every step.

`boypointer 20`

Places the boy's pointer at 20 (or other position specified).

`boyposition`

Reports, or outputs, the boy's position.

`boystep`

Reports, or outputs, the boy's step size.

`changeboystepby 1 [when time = 25]`

Adds 1 (or other specified amount) to the boy's present step size when the condition in the list (information in square brackets) is true.

`changeboystepto 2 [when girlposition = boyposition]`

Changes the boy's step size to 2 (or other specified size) when the condition in the list (in brackets) is true.

`changegirlstepby 3 [when time = 10]`

Adds 3 (or other specified amount) to the girl's present step size when the condition in the list (in brackets) is true.

`changegirlstepto -1 [girlposition = 20]`

Changes the girl's step size to –1 (or other specified size) when the condition in the list is true.

`marksevery 10`

Changes to 10 (or other specified amount) the interval at which arrows appear on the tracks below the boy and girl. The interval is initially set at 2.

`girlpointer 55`

Places the girl's pointer at 55 (or other specified position).

`girlposition`

Reports, or outputs, the girl's position.

`girlstep`

Reports, or outputs, the girl's step size.

`startboy [when girlposition = 10]`

Starts the boy when the condition in the list (in brackets) is true.

```
startboyposition 20
```

Starts the boy at 20 (or other specified position).

```
startboystep .5
```

Sets the boy's starting step size at 0.5 (or other specified size).

```
startgirl [when time = 20]
```

Starts the girl when the condition in the list (in brackets) is true.

```
startgirlposition 35
```

Starts the girl at 35 (or other specified position).

```
startgirlstep 4
```

Sets the girl's starting step size at 4 (or other specified size).

```
stop [when girlposition = 10]
```

Stops everything when the condition in the input list is true. Because both the boy and girl stop at the same time, there is only one "stop" command.

```
time
```

Reports, or outputs, the time.

***When* Lists** Many of the commands need a "*when* list" (the words shown in brackets) as an input. At the point at which the input in the *when* list is true, the command performs some action. For example:

```
stop [when boyposition = 80]
```

Note: The word *when* in these lists is optional. The same command could be expressed as follows:

```
stop [boyposition = 80]
```

In either case, the trip would stop when the boy's position is 80.

Here's another type of *when* instruction:

```
stop [when girlposition =
boyposition]
```

This trip would stop when the girl's position is equal to the boy's position. Note that this will be interesting only if the boy and girl do *not* start at the same position. Otherwise, they'll never get started. You could use a command like one of these to remedy that situation:

```
startboy [when girlposition = 10]
startboy [time = 20]
```

Either of these *when* instructions gets the boy started later than the girl.

To change the step size of the girl to 2 when the girl is at the same position as the boy, use this command:

```
changeboystepto 2 [when
girlposition = boyposition]
```

To increase the step size of the girl constantly, one change at every step, use the following:

```
changegirlstepby 1 [always]
```

Trips Messages

The boy and girl on the screen respond to commands as if they were robots. If they do not understand a command, a dialogue box may appear with one of the following messages. Read the message, click on **[OK]** or press **<return>** on the keyboard, and correct the situation as needed.

Disk or directory full

The computer disk is full.

- Use **Save My Work As...** to choose a different disk.

I don't know how to *name*.

Program does not recognize the *name* command as written. Perhaps it is misspelled. Choose **Vocabulary** under the **Help** menu to get exact command names. You can copy these names right from the help window.

I don't know what to do with *name*.

Either you gave too many inputs to a command, or no command at all.

I'm having trouble with the disk or drive.

The disk might be write-protected, there is no disk in the drive, or some similar problem.

- **Use Save My Work As...** to choose a different disk.

***Name* needs more inputs.**

Command *name* needs more inputs, such as a number or a condition list. Choose **Vocabulary** under the **Help** menu for more assistance.

***Number* is too big [too small].**

There are limits to numbers *Trips* can use.

- Don't exceed the limit.

Out of space

There is no free memory left in the computer.

- Eliminate commands you don't need.
- Save and start new work.

The maximum value for *name* is *number*.

The input is too high.

- Use a smaller number.

The minimum value for *name* is *number*.

The input is too low a number.

- Use a higher number.

Other Disk Functions

Saving Work on a Different Disk

For classroom management purposes, you might want to save student work on a disk other than the program or hard disk. Make sure that the save-to disk has been initialized (see instructions for your computer system).

1. Insert the save-to disk into the drive.
2. Choose **Save My Work As...** from the **File** menu.

The name of the disk the computer is saving to is displayed in the dialogue box. To choose a different disk, click the **[Desktop]** button and double-click to choose and open a disk from the new menu.

3. Type a name for your work if you want to give it a new or different name from the one it currently has.
4. Click on **[Save]**.

Deleting Copies of Student Work

When students no longer need previously saved work, you may want to delete their work (called *files*) from a disk. This cannot be accomplished from inside the *Trips* program. However, you can delete files from disks at any time by following directions for deleting a file for your computer system.

Trouble-Shooting

This section contains suggestions for how to correct errors, how to get back to what you want to be doing when you are somewhere else in the program, and what to do in some troubling situations. If you are new to using the computer, you might also ask a computer coordinator or an experienced friend for help.

Nothing happens after double-clicking on the *Trips* icon.

- If you are sure you double-clicked correctly, wait a bit longer. *Trips* takes a while to open or load and nothing new will appear on the screen for a few seconds.
- On the other hand, you may have double-clicked too slowly, or moved the mouse between your clicks. In that case, try again.

Wrong setting.

- Choose the correct setting from the Settings menu.

Text written in wrong area.

- Delete text.
- Click cursor in the desired area or on the desired line and retype text; or select text and use **Cut** and **Paste** from the **Edit** menu to move text to desired area.

Out of room in command window.

- Continue to enter commands. Text will scroll up and old commands will still be there, but will be temporarily out of view. To scroll, click on the up or down arrows in the scroll bar along the right side of the window.

A window closed by mistake.

- Choose **Show [window name]** from the **Windows** menu.

Windows or tools dragged to a different position by mistake.

- Drag the window back into place by following these steps: Place the pointer arrow in the stripes of the title bar. Press and hold the button as you move the mouse. An outline of the window indicates the new location. Release the button and the window moves to that location.

I clicked somewhere and now *Trips* is gone! What happened?

You probably clicked in a part of the screen not used by *Trips*, and the computer therefore took you to another application such as the Finder's desktop.

- Click on a *Trips* window, if visible.
- Double-click on *Trips* again, or select it from the application menu at the right of the menu bar.

How do I select a section of text?

In certain situations, you may wish to copy or delete a section or block of text.

- Point and click at one end of the text. Drag the mouse by holding down the mouse button as you move to the other end of the text. Release the mouse button. Use the **Edit** menu to **Copy**, **Cut**, and **Paste**, or press **[delete]**.

System Error Message

Some difficulty with the *Trips* program or your computer caused the computer to stop functioning.

- Turn off the computer and repeat the steps to turn it on and start *Trips* again. Any work that you saved will still be available to open from your disk.

I tried to print and nothing happened.

- Check that the printer is connected and turned on.
- When printers are not functioning properly, a system error may occur that causes the computer to freeze. If there is no response from the keyboard or when moving or clicking with the mouse, you may have to turn off the computer and start over. See System Error Message above.

I tried to print the *Trips* window [or the Graph window] and not everything printed.

- Choose the Color/Grayscale option for printing.
- If your printer has no such option (for example, if it is an older black-and-white printer), you need to find a different printer to print graphics.

Blackline Masters

_______________________, 19_____

Dear Family,

Our class has begun a new mathematics unit called *Patterns of Change*. We are talking about how changes can be described with tables and graphs. The children first consider tile designs that keep getting bigger in a regular way. They tell the story of each design with a number table and graphs that reflect the changing pattern.

Next we begin talking about changes in speed. The children act out "trips" by walking along a straight track of tape on the floor, changing their speed as they go. Again, the children make tables and graphs to show the changing speeds. In similar work on the computer, the children plan and run "trips" for two walkers. They make changes in the walkers' rate of speed, in their starting position, and in their direction of movement, and they try to predict the outcome as they explore the effects of these changes on "who gets there first."

To learn more about what we're doing, ask your child to tell you about the activities that he or she brings home.

- Look at your child's drawings of tile patterns, and at the tables and graphs that go with them. Ask your child to explain the patterns to you. How can you tell how the design will continue to grow? With your child, design some other tile patterns that grow in predictable ways.
- See if you can interpret the diagram your child makes to show how the speed changes during a walk along a straight line. Can you tell the story of the trip? It might be something like this: "Start out walking slowly. Stop halfway for 6 seconds. Then run to the end."
- Over the next few weeks, help your child look for things that change in different ways and at different speeds. Can you find some things that change faster and faster? Can you find things that change steadily? Can you find anything that changes by gradually slowing down? by gradually shrinking?
- Look in newspapers and other print material for graphs and tables that show something changing over time. Work with your child to make sense of these.

Making and interpreting tables and graphs that show change over time is an important skill. The understandings that our class gains in this unit will help in their future studies in both math and science.

Sincerely,

Name Date

Student Sheet 1

Tile Pattern Template

Growing Tile Pattern

Step number	New tiles (step size)	Total so far
1		
2		
3		
4		
5		
6		
7		
8		

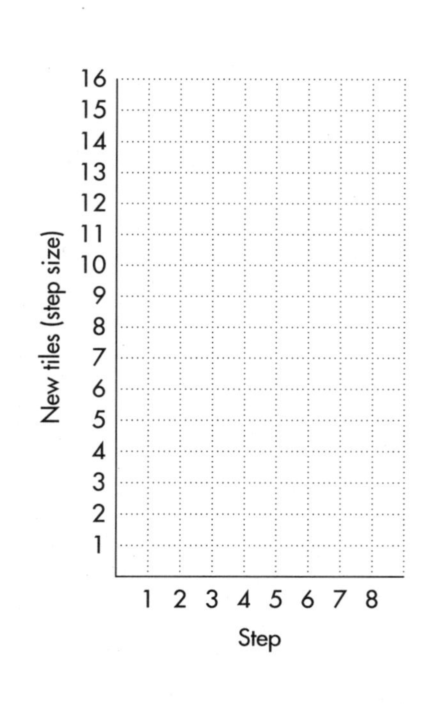

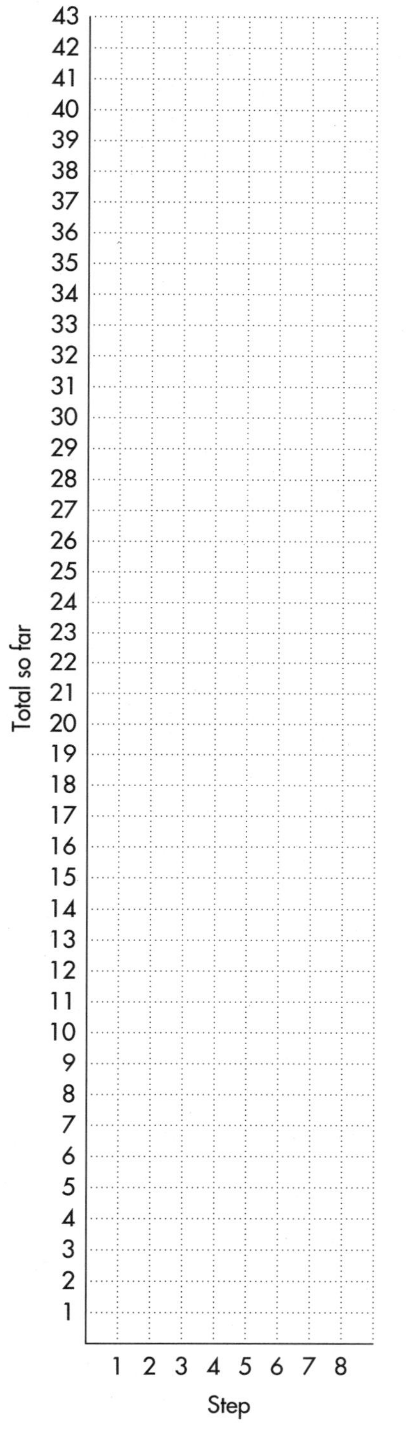

Name Date

Student Sheet 2

Growing Tile Patterns (page 1 of 4)

Twos Tower

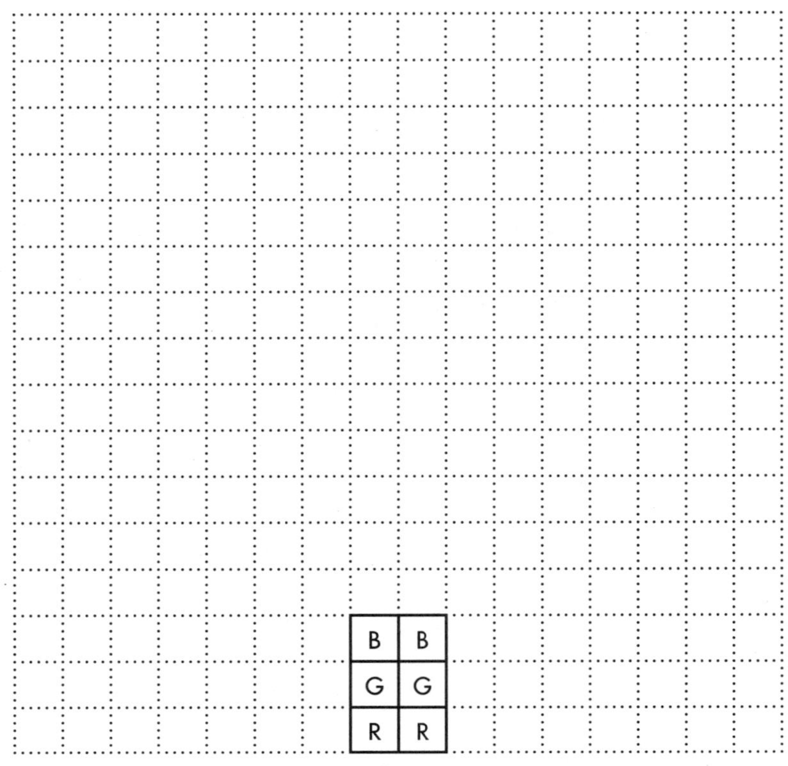

Step number	New tiles (step size)	Total so far
1	2	2
2	2	4
3	2	6
4	2	
5	2	
6		
7		
8		

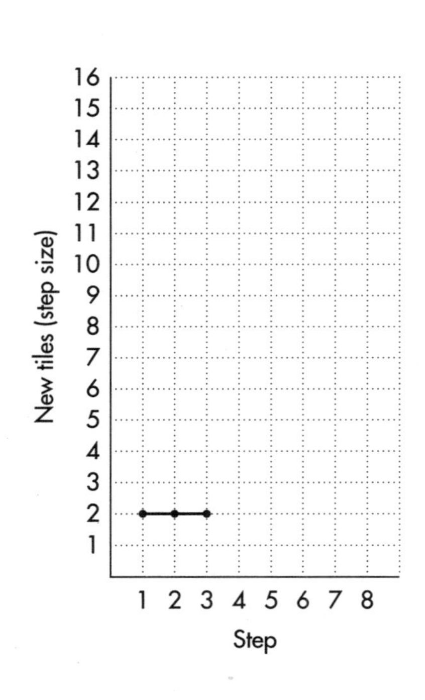

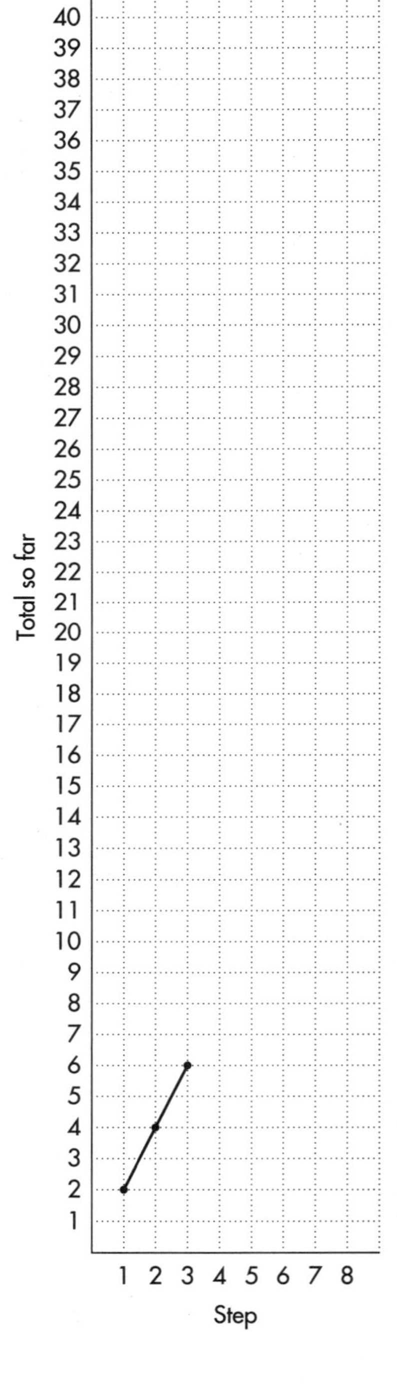

Name Date

Growing Tile Patterns (page 2 of 4)

Squares

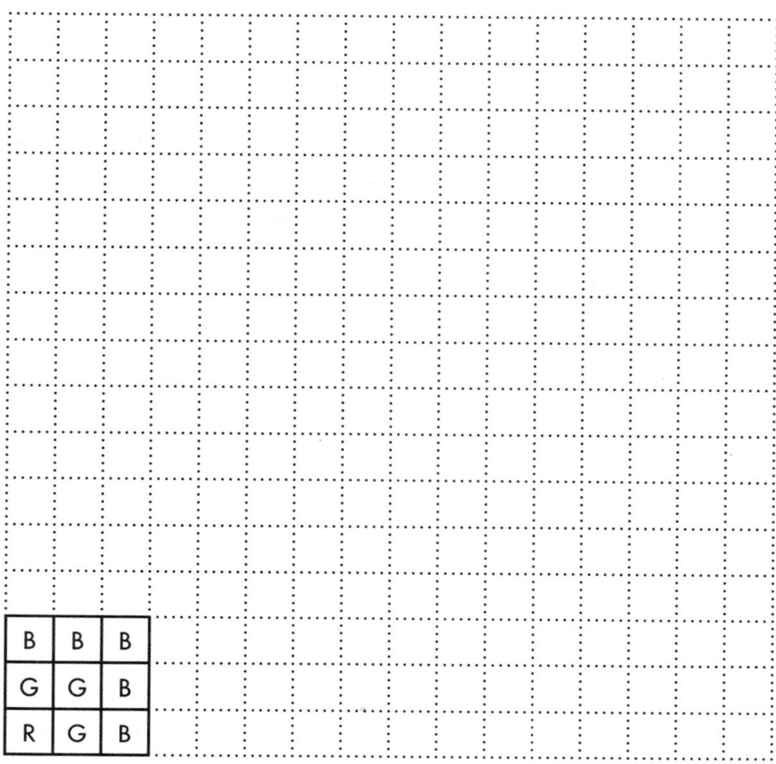

Step number	New tiles (step size)	Total so far
1	1	1
2	3	4
3	5	9
4		
5		
6		
7		
8		
0		

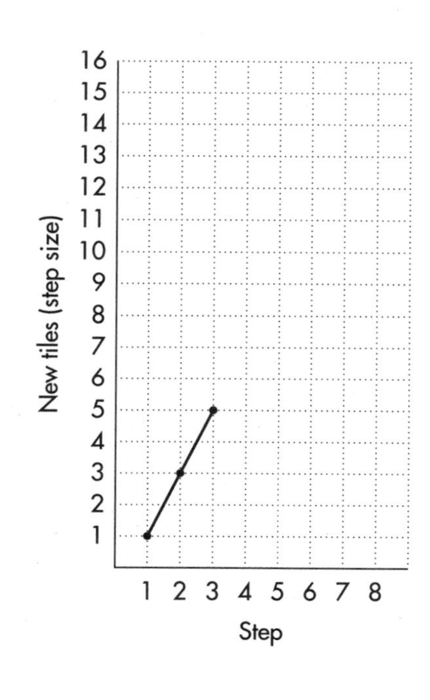

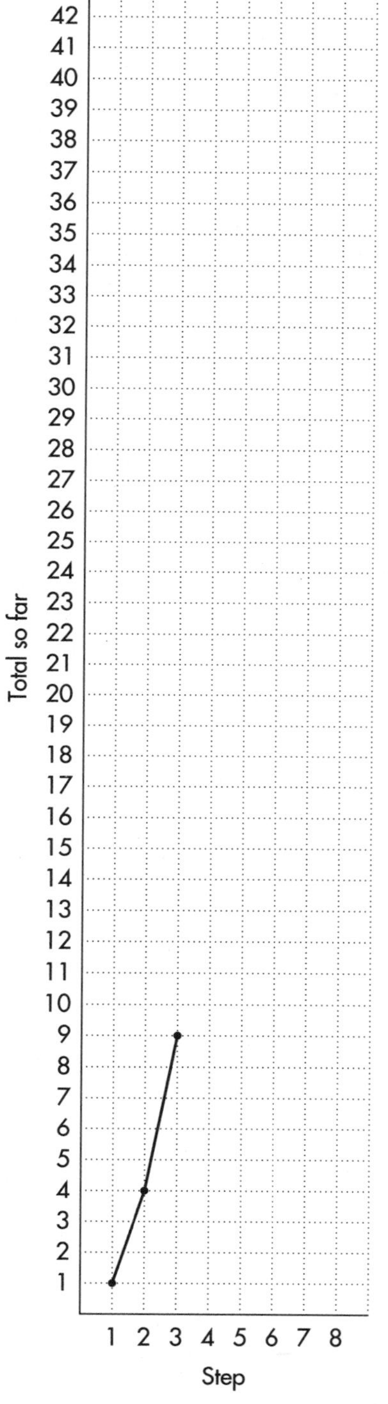

Name Date

Student Sheet 2

Growing Tile Patterns (page 3 of 4)

Staircase

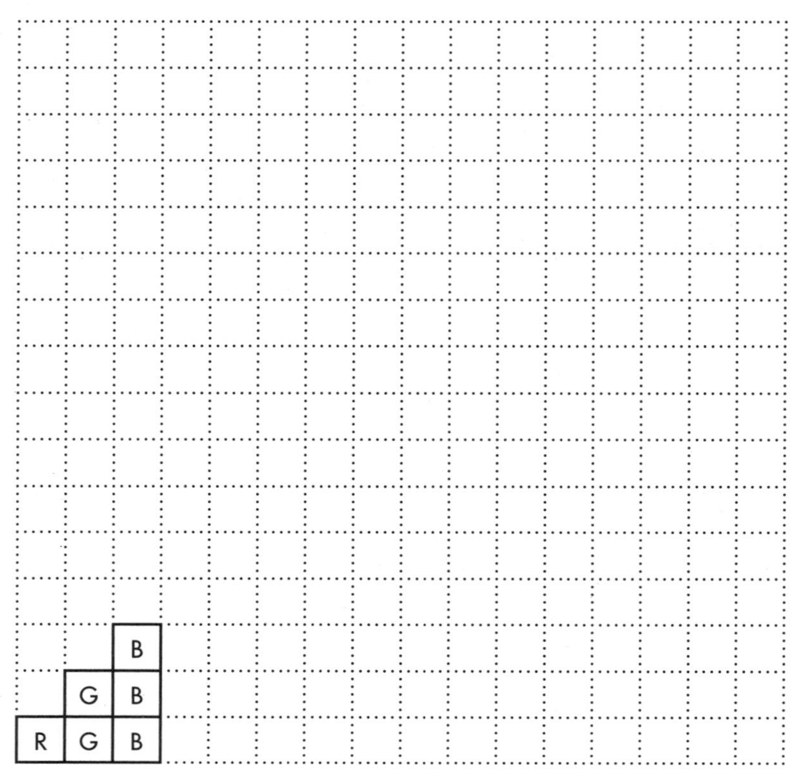

Step number	New tiles (step size)	Total so far
1	1	1
2	2	3
3	3	6
4		
5		
6		
7		
8		

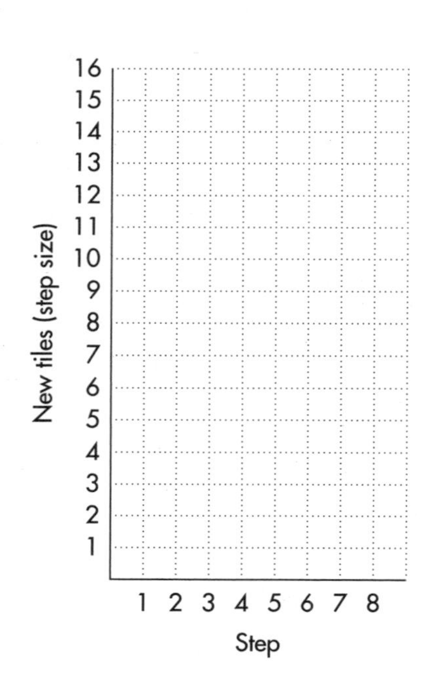

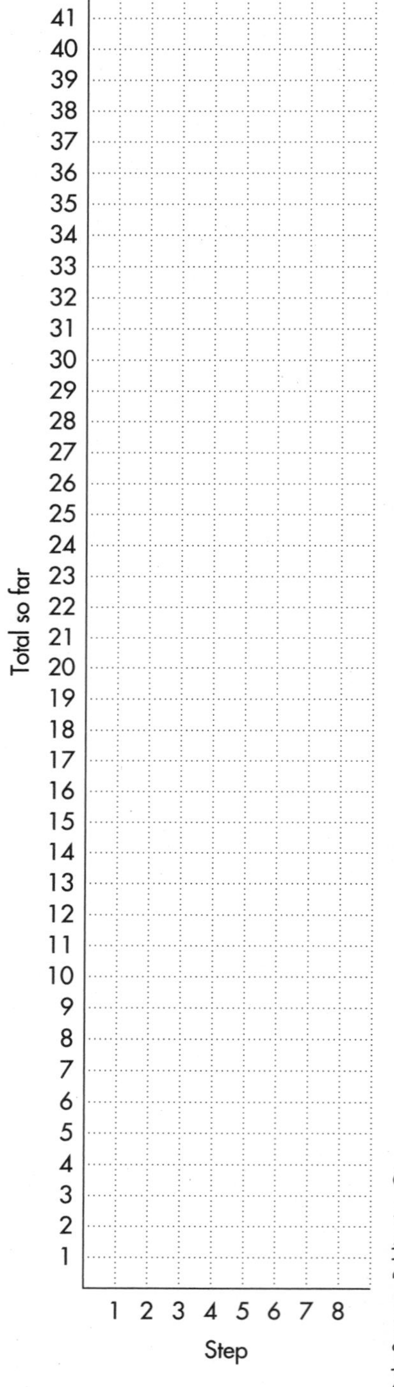

Name Date

Growing Tile Patterns (page 4 of 4)

Doubling

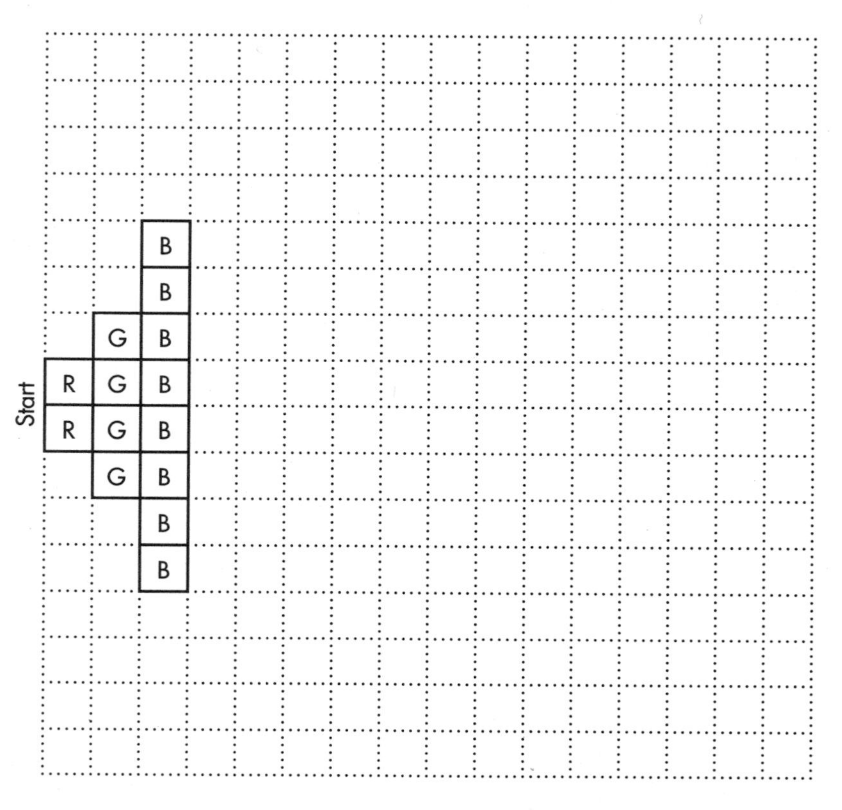

Step number	New tiles (step size)	Total so far
1	2	2
2	4	6
3	8	14
4	16	30
5		
6		
7		
8		

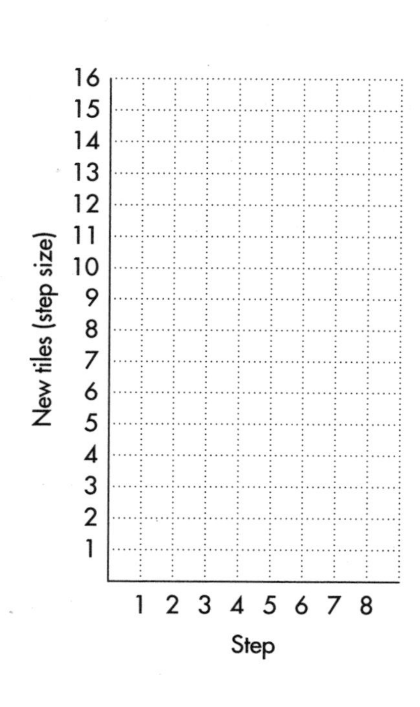

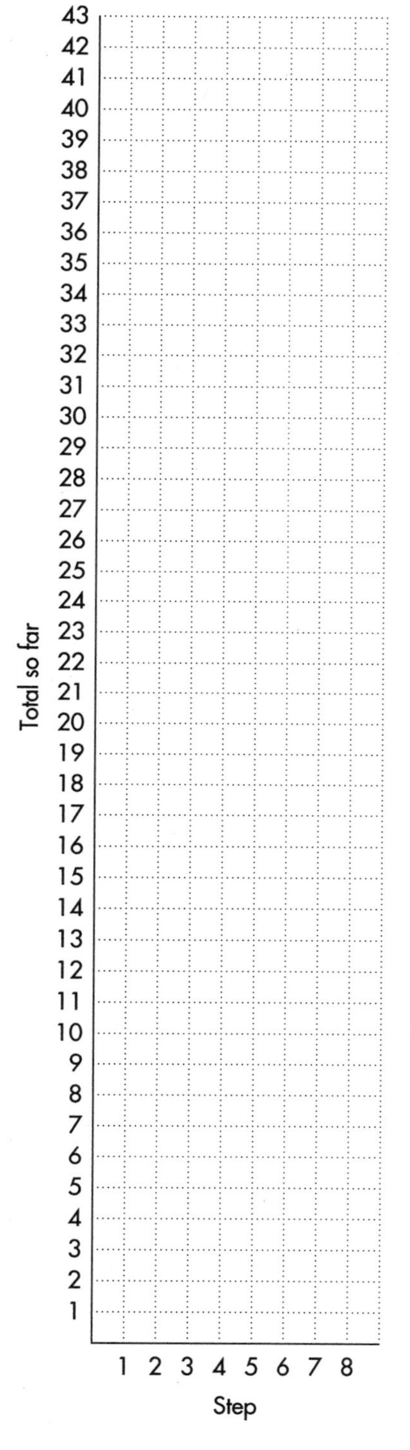

Name Date

Student Sheet 3

Template for Tables

Time

Time (seconds)	Total distance so far (meters)
2	
4	
6	
8	
10	
12	
14	

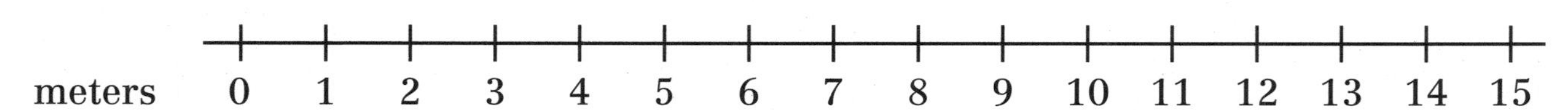

Time (seconds)	Total distance so far (meters)
2	
4	
6	
8	
10	
12	
14	

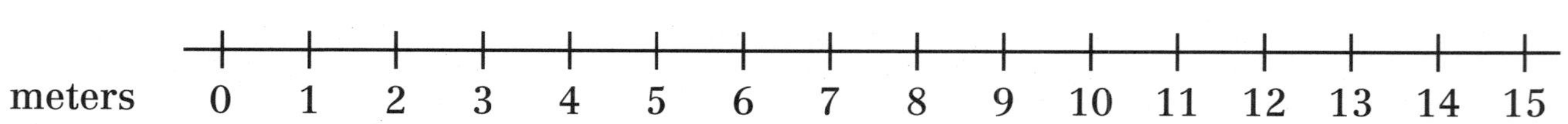

Name Date

Height of a Girl

This table lists the heights of a girl from age 7 to 17.

Make a graph of her changing height.

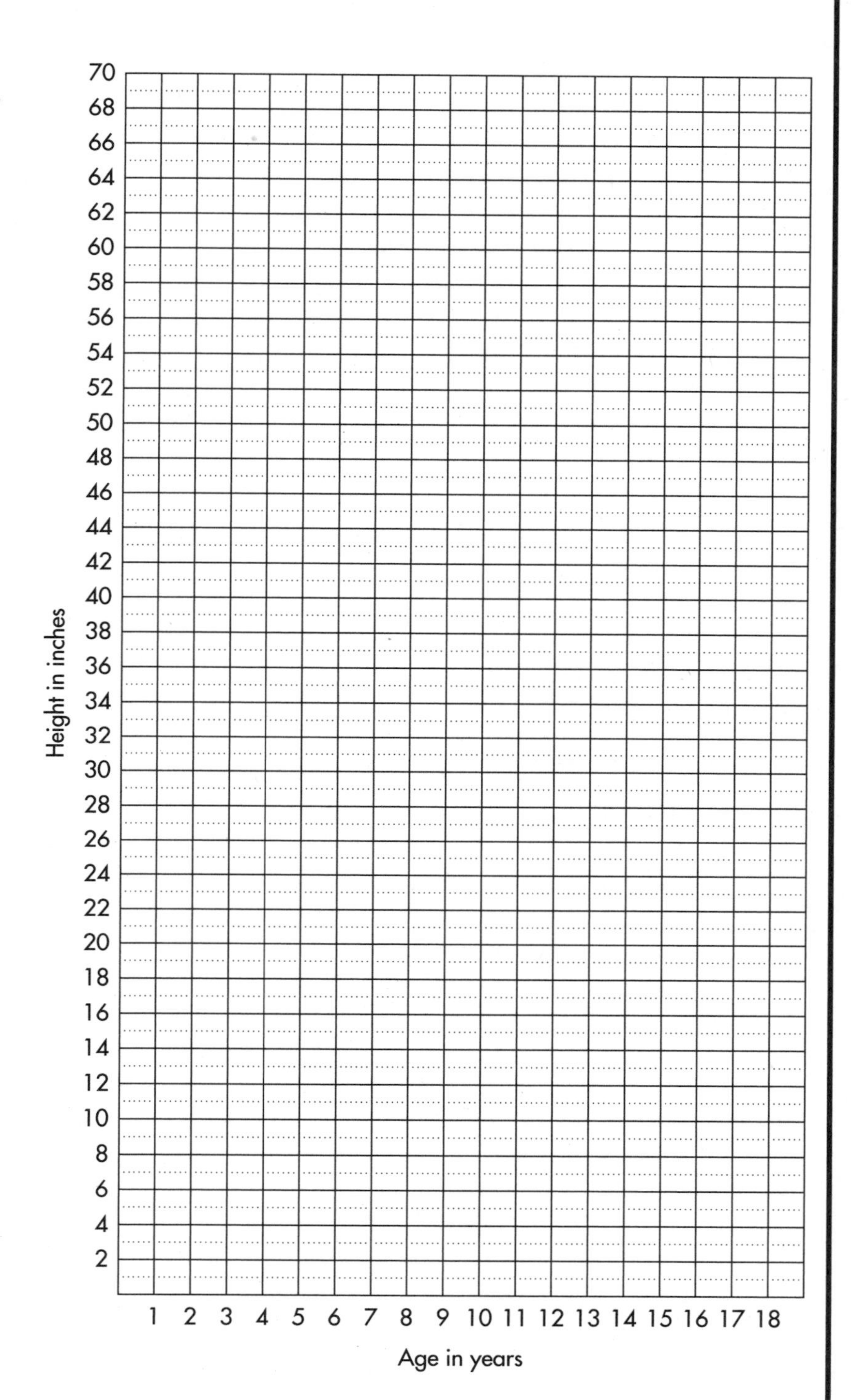

Age (years)	Height (inches)	Growth in last year
7	48	
8	50	
9	52	
10	54	
11	56	
12	59	
13	62	
14	65	
15	66	
16	67	
17	67	

Tell the story of the girl's growth. When did she grow fast? When did she grow slowly? What is happening to her growth at the end?

Name Date

Student Sheet 5

Three Motion Stories

Plan a trip for one of the following stories. Make a table showing where the person would be every 2 seconds (that is, where the beanbags would land).

Story A Run a few steps, stop, run a few steps, stop, then walk to the end.

Story B Walk very slowly a short way, stop for about 6 seconds, and then walk fast to the end.

Story C Run about halfway, then go slower and slower until the end.

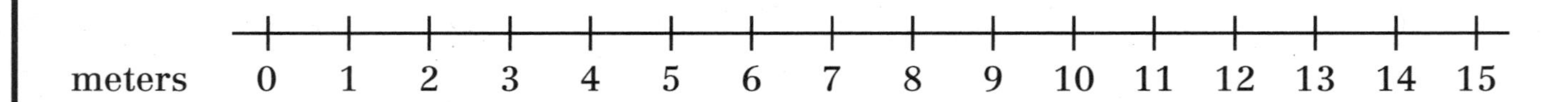

Time (seconds)	Total distance so far (meters)
2	
4	
6	
8	
10	
12	
14	

Name Date

Student Sheet 6

Graph of a Trip

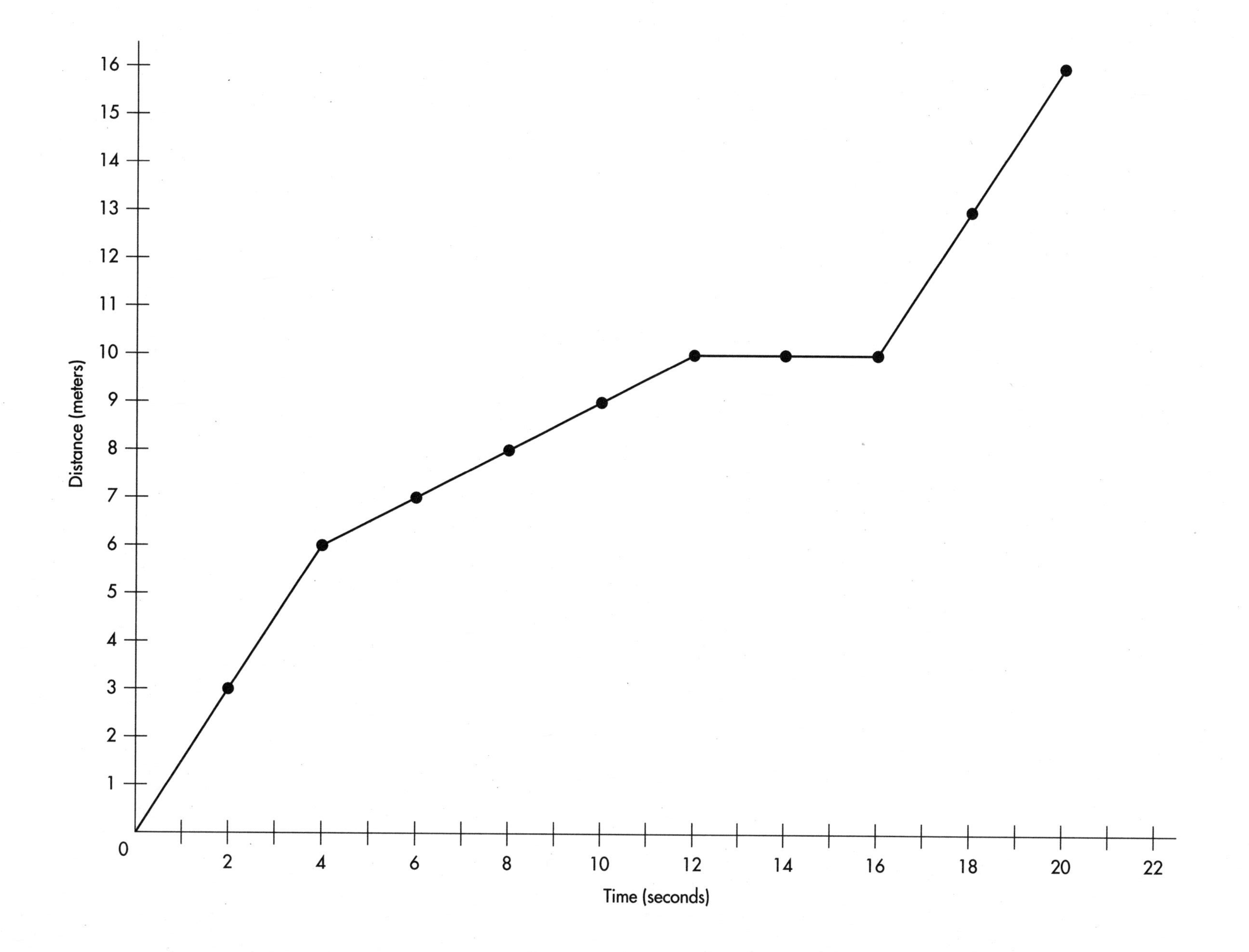

Name Date

Student Sheet 7

Graph Template

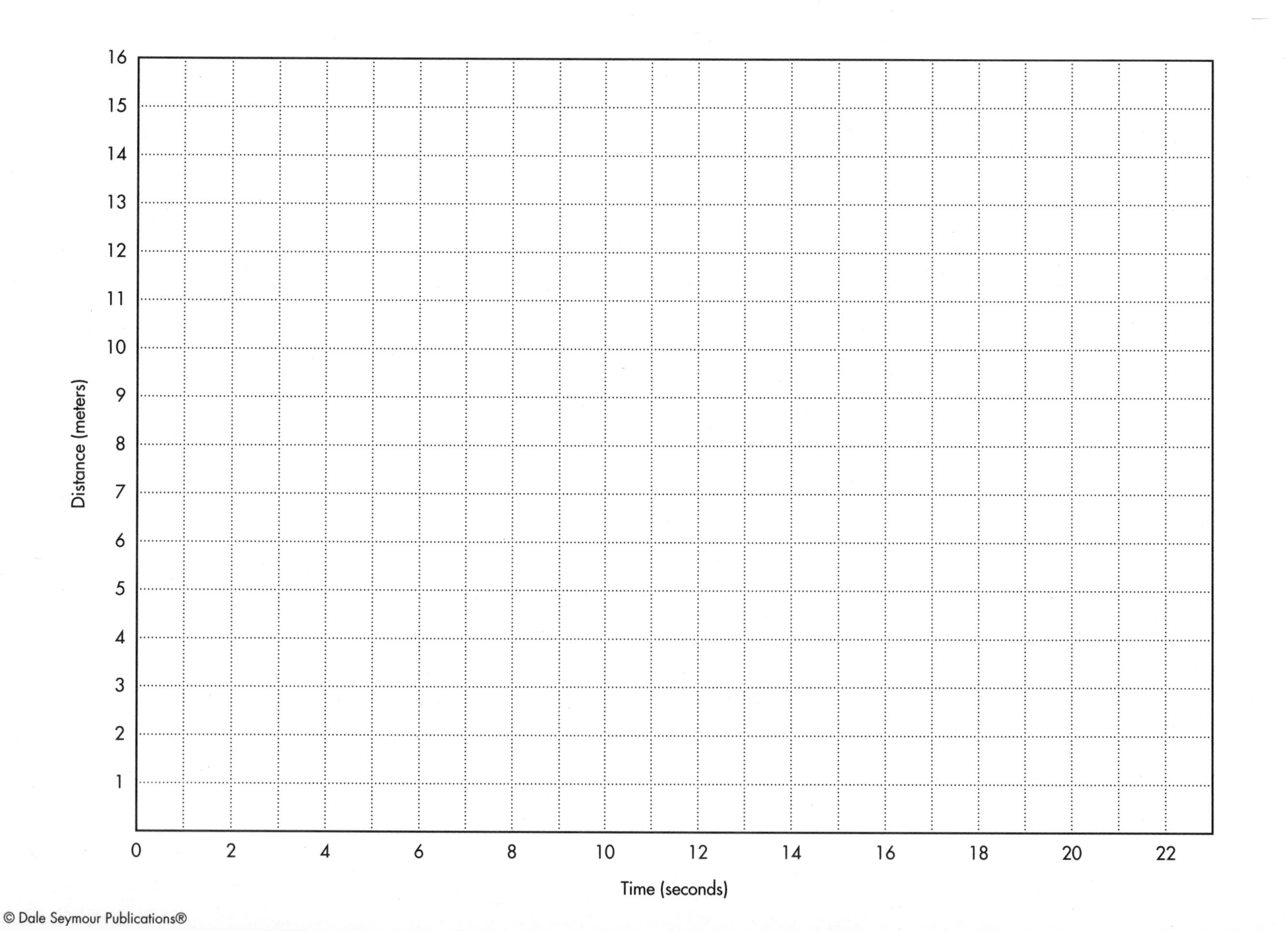

Name Date

Student Sheet 8

Matching Stories, Tables, and Graphs (page 1 of 2)

Cut apart the stories, the tables, and the graphs.
Which ones match? Group them together.
Finish filling in the tables, and write the missing story.

Story 1 Walk slowly about halfway and then run until the end.
Story 2 Run about halfway, stop for 4 seconds, then walk to the end.
Story 3

Table A

Time (seconds)	Distance in previous 2 seconds (step size)	Total distance (meters)
2	1	1
4	1	2
6	1	3
8		4
10		5
11		8
12		11
14		

Table B

Time (seconds)	Distance in previous 2 seconds (step size)	Total distance (meters)
2	2	2
4	2	4
6	2	6
8		7
10		5
12		3
14		1
16		

Name Date

Student Sheet 8

Matching Stories, Tables, and Graphs (page 2 of 2)

Table C

Time (seconds)	Distance in previous 2 seconds (step size)	Total distance (meters)
2		
4		
6		
8		
10		
12		
14		

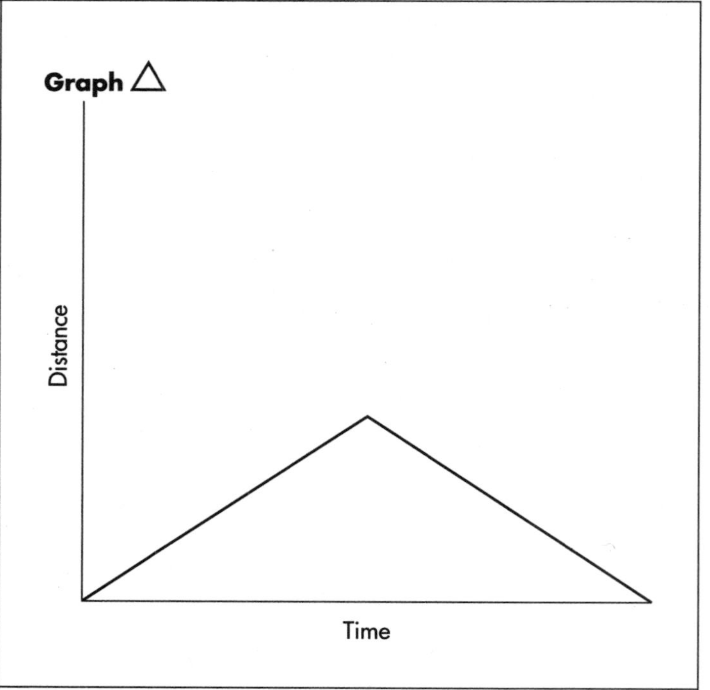

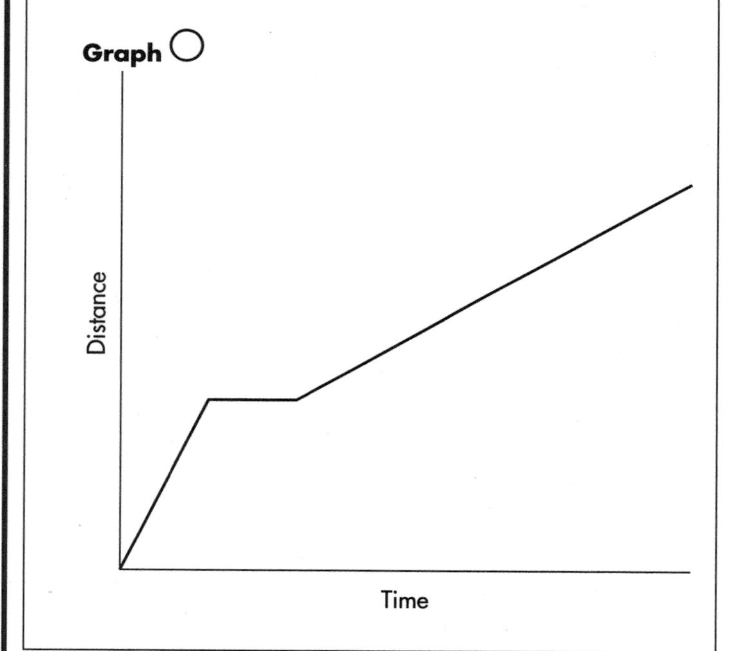

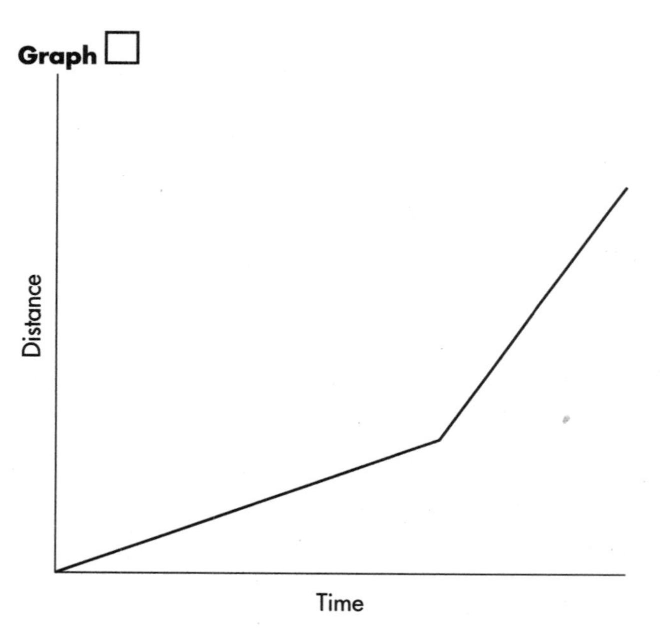

Name

Date

Student Sheet 9

Trips Computer Screen

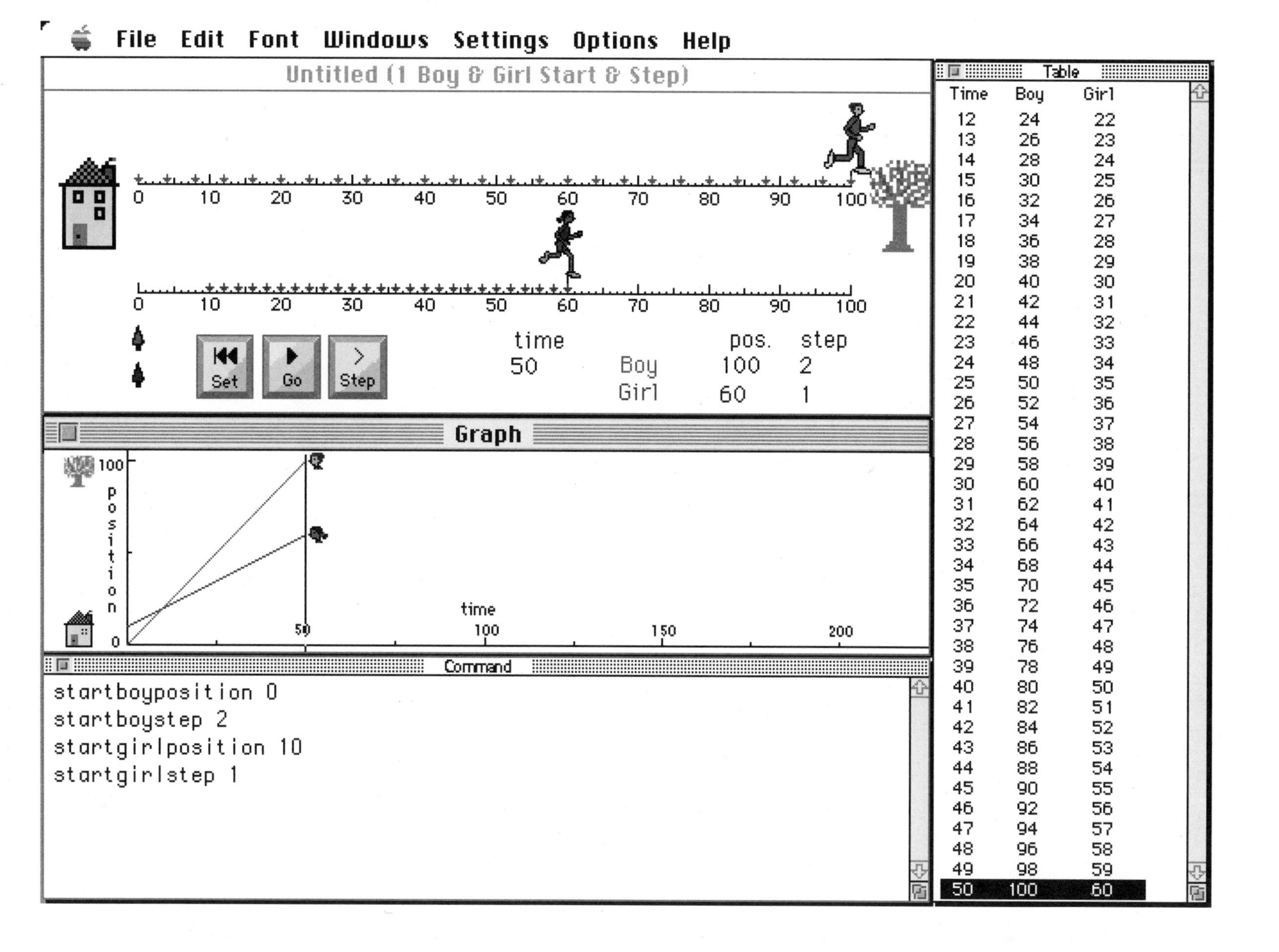

Name Date

Student Sheet 10

Trips in Setting 1 (page 1 of 2)

Create a trip for Motion Stories 1, 2, and 3. Use either the meterstick or the computer. In the blanks, write the values for position and step size that you used for each trip.

Note: If you have done a story with the meterstick, check it on the computer if you have time.

Motion Story 1 The girl gets to the tree way ahead of the boy.

startboyposition _____ *(The boy is going to start at position ___?___)*

startboystep _____ *(The boy is going to walk with a step size of ___?___)*

startgirlposition _____ *(The girl is going to start at position ___?___)*

startgirlstep _____ *(The girl is going to walk with a step size of ___?___)*

Motion Story 2 The girl starts behind the boy, but she passes the boy and gets to the tree first.

startboyposition _____

startboystep _____

startgirlposition _____

startgirlstep _____

Motion Story 3 The boy starts at the tree and the girl starts at the house. The boy gets to the house before the girl gets to the tree.

startboyposition _____

startboystep _____

startgirlposition _____

startgirlstep _____

Choose the trip for one story and use it to fill in the next page. Record here which story you choose: ________________

Name Date

Trips in Setting 1 (page 2 of 2)

Tell how the trip would look from the point of view of either the boy or the girl.

Time	Position of boy	Position of girl
0		

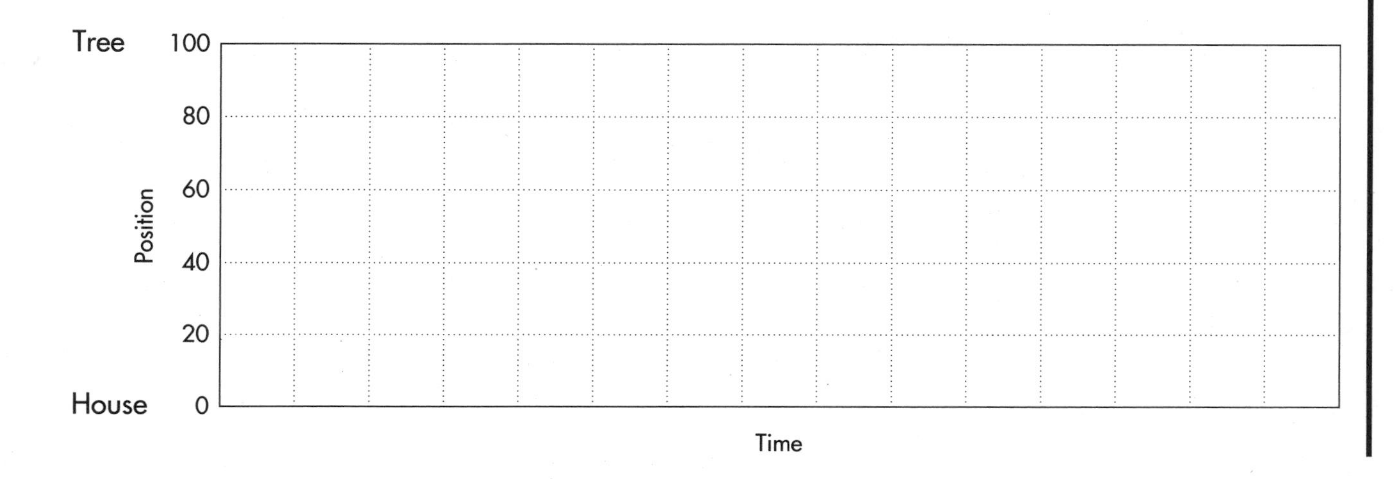

Name Date

Story of a Trip (page 1 of 2)

Story

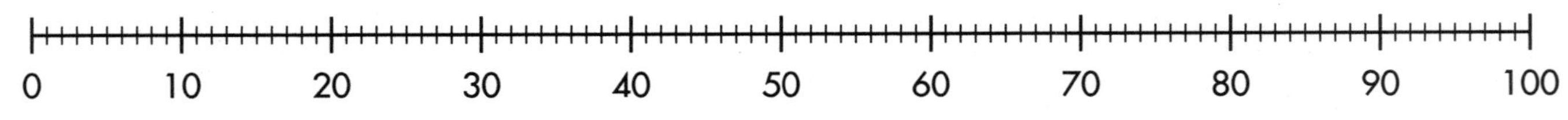

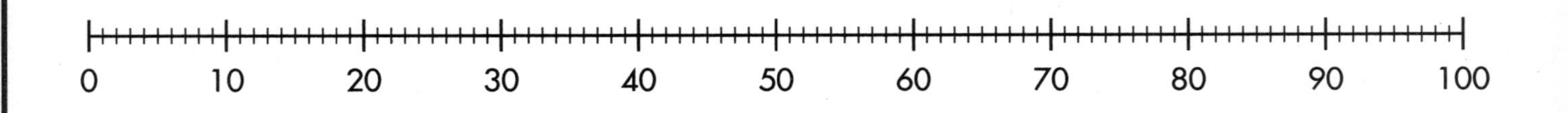

Time (seconds)	Position of boy (meters)

Time (seconds)	Position of boy (meters)

Name Date

Story of a Trip (page 2 of 2)

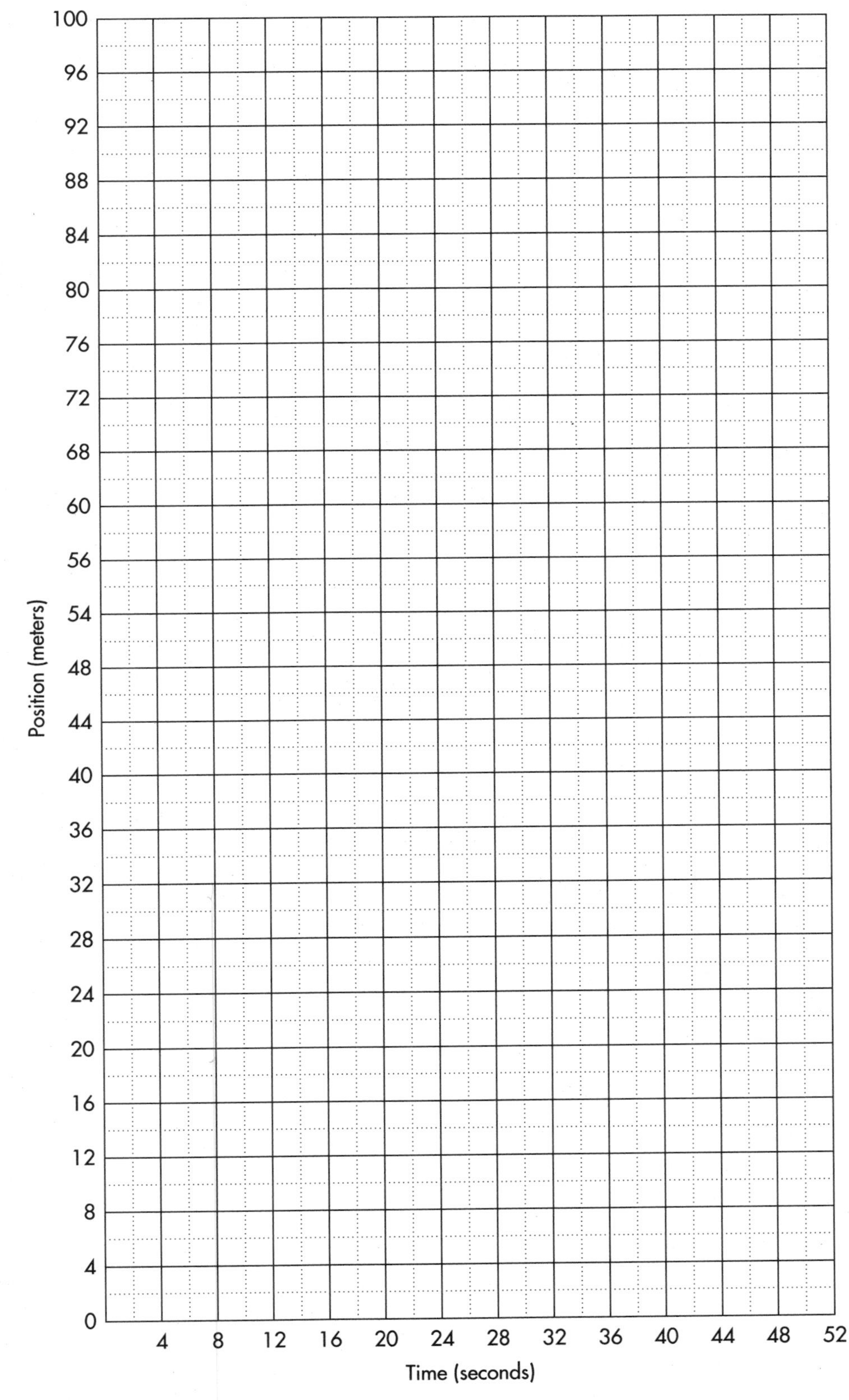

Name Date

Student Sheet 12

Using the *Trips* Settings

Setting 1

```
startboyposition 50
startboystep -1
startgirlposition 0
startgirlstep 3
```

	Boy		Girl	
Time	Step size	Position	Step size	Position
0	–1	50	3	0
1	–1	49	3	3
2	–1	48	3	6
3		47		9
4				
5				
6				
7				

Setting 2

```
startboyposition 0
startboystep 2
startgirlposition 0
startgirlstep 4
changeboystepto 8
 [when boyposition = 4]
changegirlstepto 6
 [when girlposition = 12]
```

	Boy		Girl	
Time	Step size	Position	Step size	Position
0	2	0	4	0
1	2	2	4	4
2	2	4	4	8
3	8	12	4	12
4	8	20	6	18
5		28		24
6				
7				

Setting 3

```
startboyposition 0
startboystep 0
startgirlposition 0
startgirlstep 10
changeboystepby 1 [always]
changegirlstepby -1
[always]
```

	Boy		Girl	
Time	Step size	Position	Step size	Position
0	0	0	10	0
1	1	0	9	10
2	2	1	8	19
3	3	3		
4	4	6		
5				
6				
7				

Name Date

Student Sheet 13

Trips in Setting 2 (page 1 of 2)

Create a trip for Motion Stories 1, 2, and 3. In the blanks, write the values for position and step size.

Note: If you have done a story with the meterstick, check it on the computer if you have time.

Motion Story 1 The girl gets to the tree way ahead of the boy.

startboyposition ____ startboystep ____

startgirlposition ____ startgirlstep ____

changeboystepto ____ [when boyposition = ____]

(The boy changes his step size to ? when he reaches position ?)

changegirlstepto ____ [when girlposition = ____]

(The girl changes her step size to ? when she reaches position ?)

Motion Story 2 The girl starts behind the boy, but she passes the boy and gets to the tree first.

startboyposition ____ startboystep ____

startgirlposition ____ startgirlstep ____

changeboystepto ____ [when boyposition = ____]

changegirlstepto ____ [when girlposition = ____]

Motion Story 3 The boy starts at the tree and the girl starts at the house. The boy gets to the house *before* the girl gets to the tree.

startboyposition ____ startboystep ____

startgirlposition ____ startgirlstep ____

changeboystepto ____ [when boyposition = ____]

changegirlstepto ____ [when girlposition = ____]

Choose the trip for one story and use it to fill in the next page. Record here which story you choose: ________________

Name Date

Student Sheet 13

Trips in Setting 2 (page 2 of 2)

Tell how the trip would look from the point of view of either the boy or the girl.

Time	Position of boy	Position of girl
0		

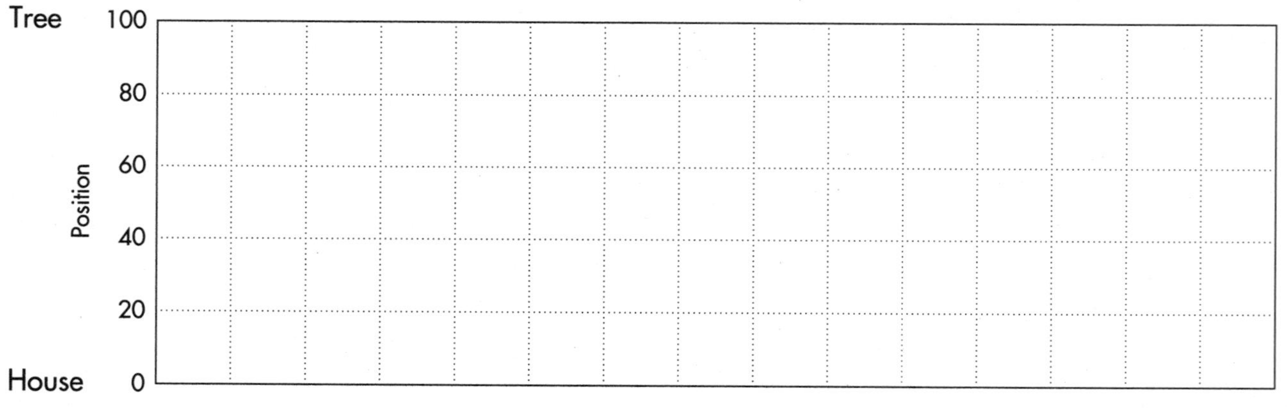

Name

Date

Student Sheet 14

Trips in Setting 3 (page 1 of 2)

Create a trip for Motion Stories 1, 2, and 3. In the blanks, write the values for position and step size.

Motion Story 1 The girl gets to the tree way ahead of the boy.

startboyposition ____ startboystep ____

startgirlposition ____ startgirlstep ____

changeboystepby ____ [always]

(At every step, the boy changes his step size by _?_)

changegirlstepto____ [always]

(At every step, the girl changes her step size by _?_)

Motion Story 2 The girl starts behind the boy, but she passes the boy and gets to the tree first.

startboyposition ____ startboystep ____

startgirlposition ____ startgirlstep ____

changeboystepto ____ [always]

changegirlstepto ____ [always]

Motion Story 3 The boy starts at the tree and the girl starts at the house. The boy gets to the house before the girl gets to the tree.

startboyposition ____ startboystep ____

startgirlposition ____ startgirlstep ____

changeboystepto ____ [always]

changegirlstepto ____ [always]

Choose the trip for one story and use it to fill in the next page. Record here which story you choose: ____________

Name

Date

Trips in Setting 3 (page 2 of 2)

Tell how the trip would look from the point of view of either the boy or the girl.

Time	Position of boy	Position of girl
0		

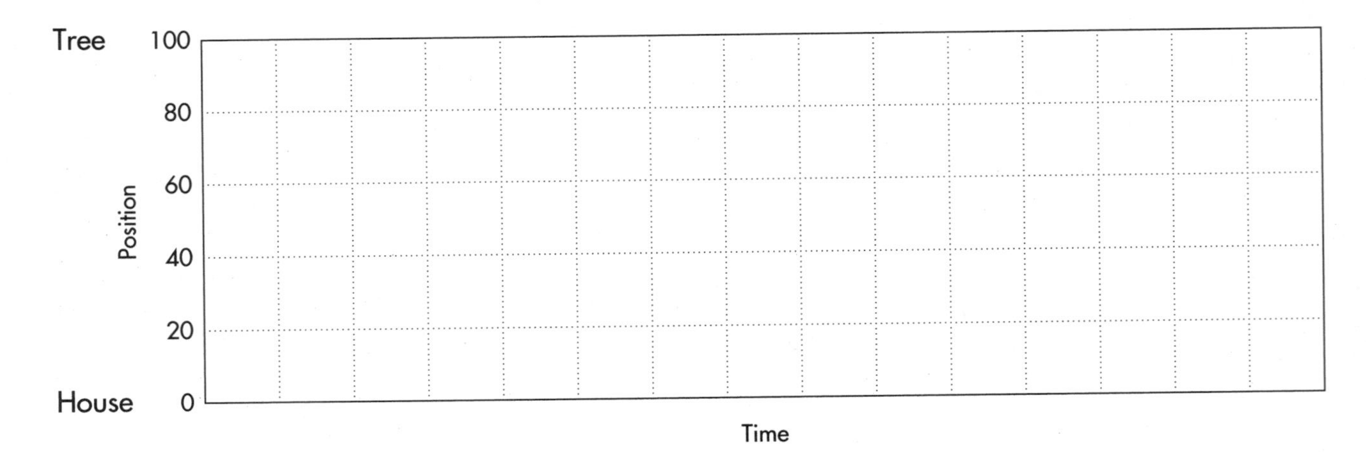

Name Date

Student Sheet 15

Two Kinds of Graphs (page 1 of 2)

Time (seconds)	Position	Step size

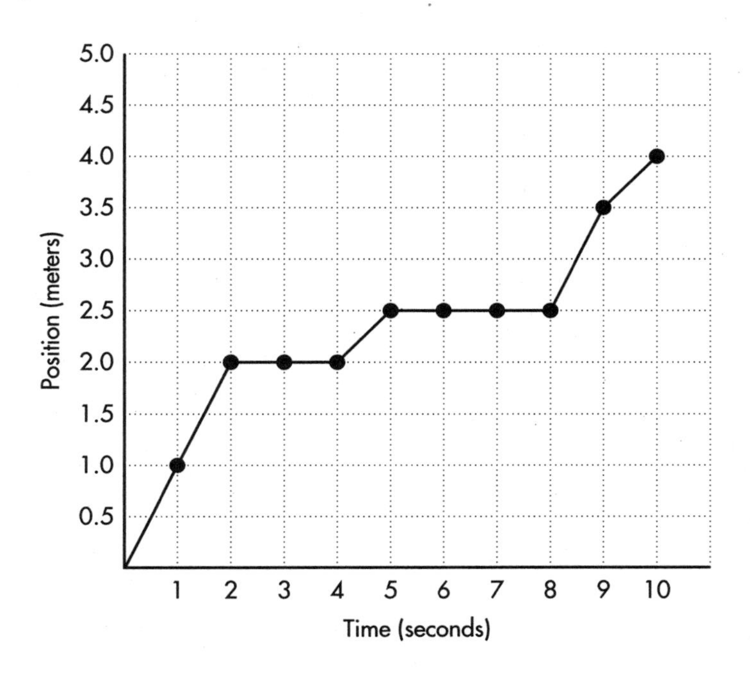

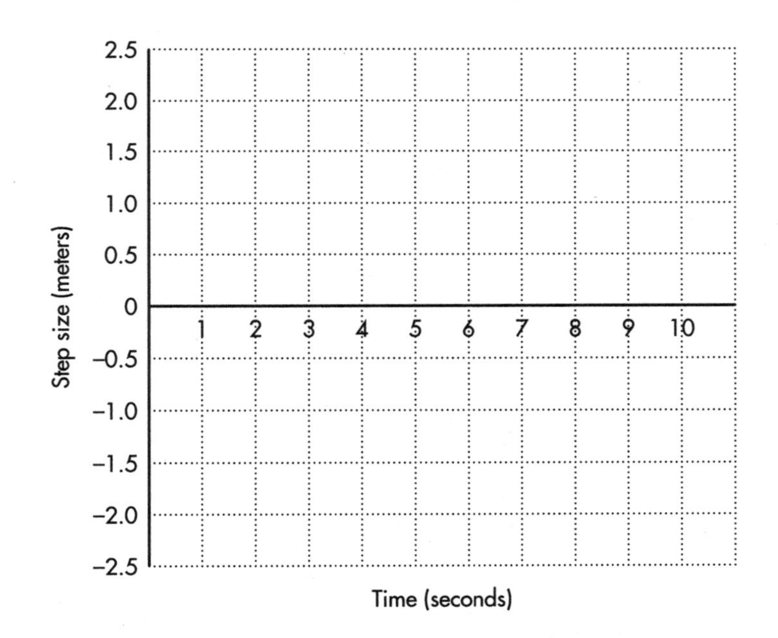

Name Date

Student Sheet 15

Two Kinds of Graphs (page 2 of 2)

Time (seconds)	Position	Step size

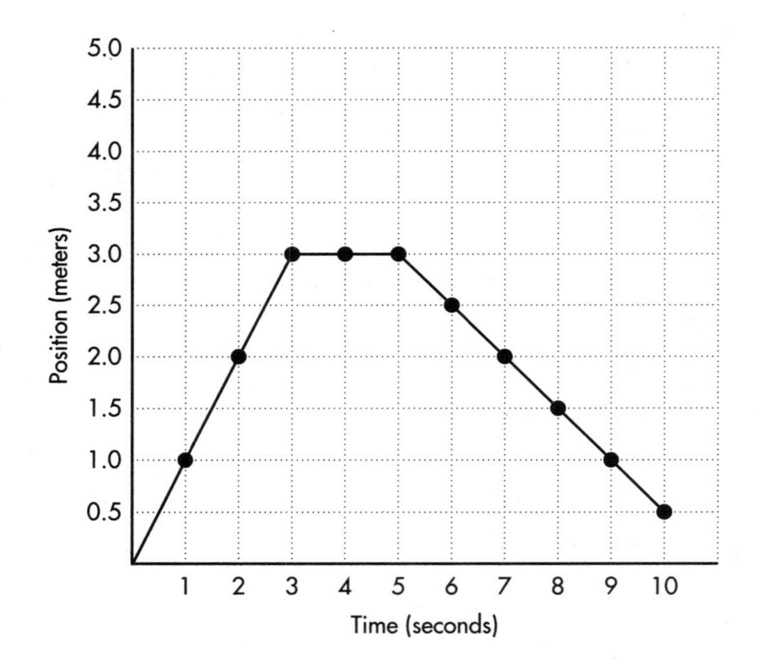

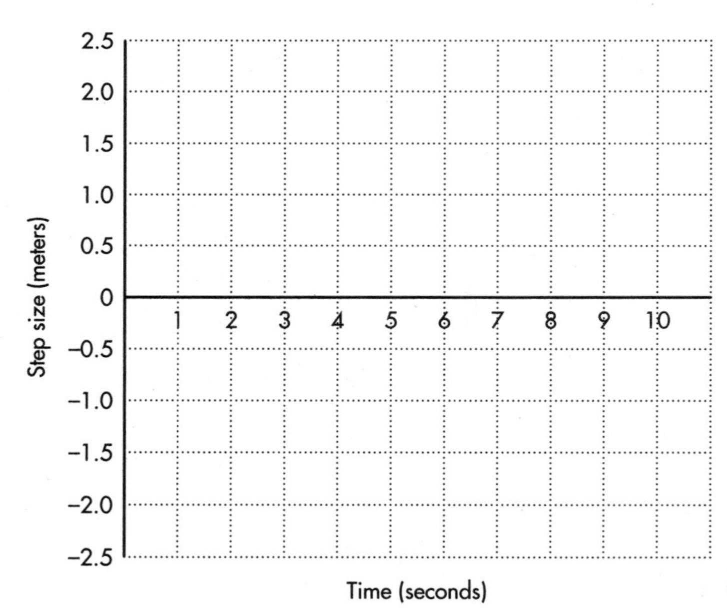

Name Date

Mystery Walks (page 1 of 2)

1. She walked slowly for 3 seconds. Then she stood still for 4 seconds. Suddenly, during the last 3 seconds, she went quite fast.

2. She ran fast for 3 seconds, then slowly for 4 seconds. Then she went back to the beginning in 5 seconds.

3. He waited for 4 seconds before starting to walk slowly with a step size of 0.5. He walked for a few seconds and then stopped.

4. He walked backward very slowly. After 5 seconds he ran forward for 5 more seconds.

5. She left home running really fast with step size of 1.5 meters. She went at that rate for 3 seconds, but then she realized that she had forgotten her book. She stopped for a couple of seconds to decide what to do. Then she decided that it would be too late anyway, so she went back home slowly.

6. From his house to the corner store is 10 meters. He ran to the store, spent 1 second looking at the CLOSED sign, and walked slowly back to his house.

7. She decided to cross the park walking slowly at first but going faster and faster each step. It took her 5 seconds to get to the other side.

8. He was going home, not in a rush. As he stepped into the street, he realized that a car was coming. He waited for the car, then ran across the street. As soon as he got to the other side of the street, he walked slowly again.

Name Date

Student Sheet 16

Mystery Walks (page 2 of 2)

9. At first the old man walked very slowly, as if he was tired. Suddenly, when he was next to us, he started to run amazingly fast. After a few seconds he stopped and walked back to say, "I surprised you, didn't I?"

10. The dog ran off to catch the stick that his owner had thrown. As the dog grabbed the stick, he saw a rabbit. The dog held very still for a moment. Then, instead of running back to his owner, he crept very slowly toward the rabbit. When the dog was close to the rabbit, he jumped forward at great speed.

11. First she went fast, at a steady pace. Then, at around 5 meters, she started to slow down. She went slower and slower until she stopped. She stood still for 4 seconds. Finally she walked slowly and steadily for a while.

12. Trying not to wake anyone up, she walked very slowly with small steps. Once she got to the door, she began to run faster and faster. After 3 seconds of running, she stopped and sat down.

13. Imagine someone walking back and forth two times between the chalkboard and her desk. She always walks fast toward the board and slowly toward the desk. At the end she remains still for 3 seconds.

14. He waited for 4 seconds before starting to run with a step size of 1.5. He ran for a few seconds and then stopped.

15. It is 8 meters between her bedroom and the kitchen. She walked into the kitchen slowly because she was half asleep, and then just stared at the room for a moment. Then she went back to bed very quickly.

Name Date

Position vs. Time Graph

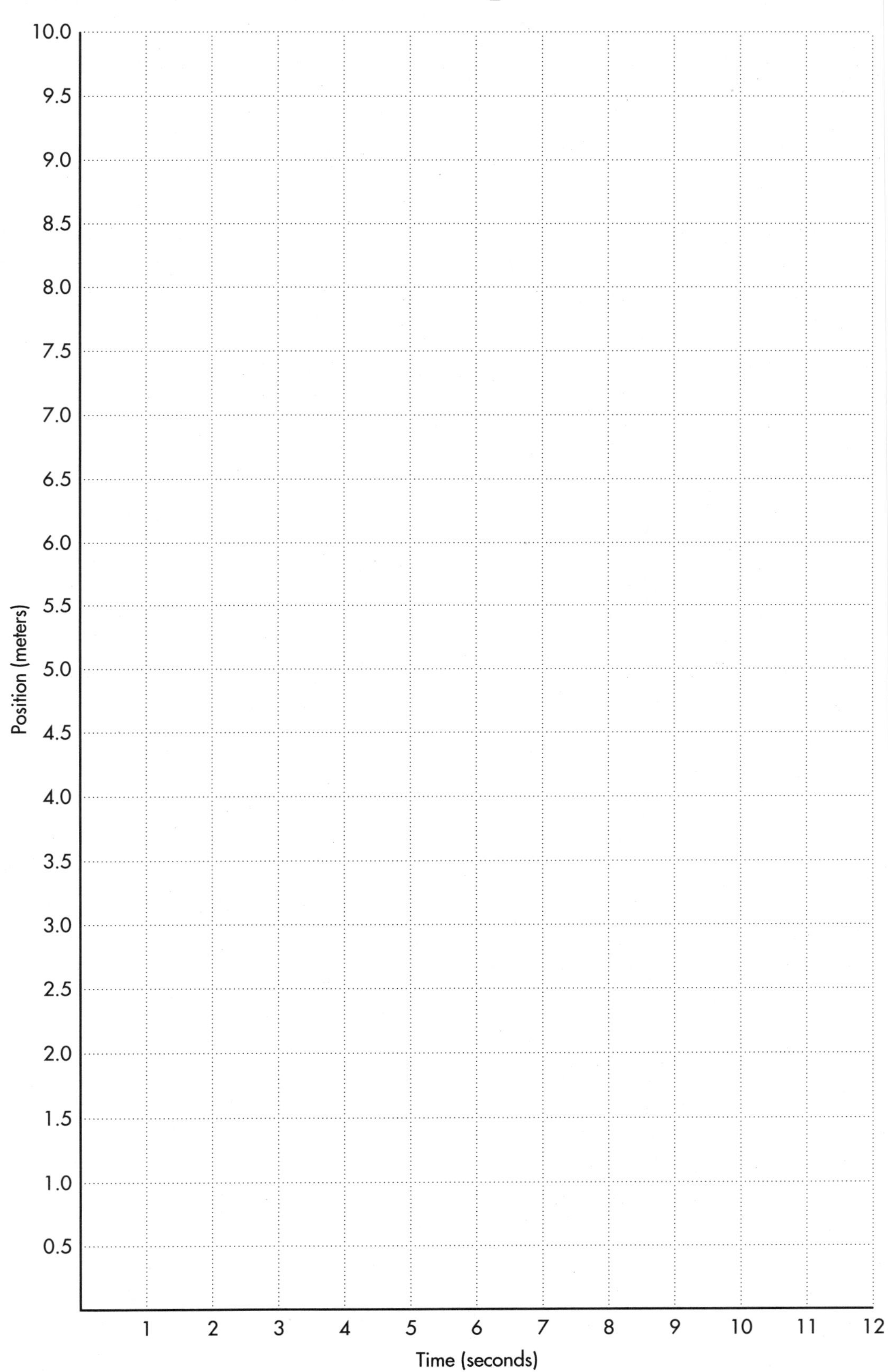

Name Date

Student Sheet 18

Step Size vs. Time Graph

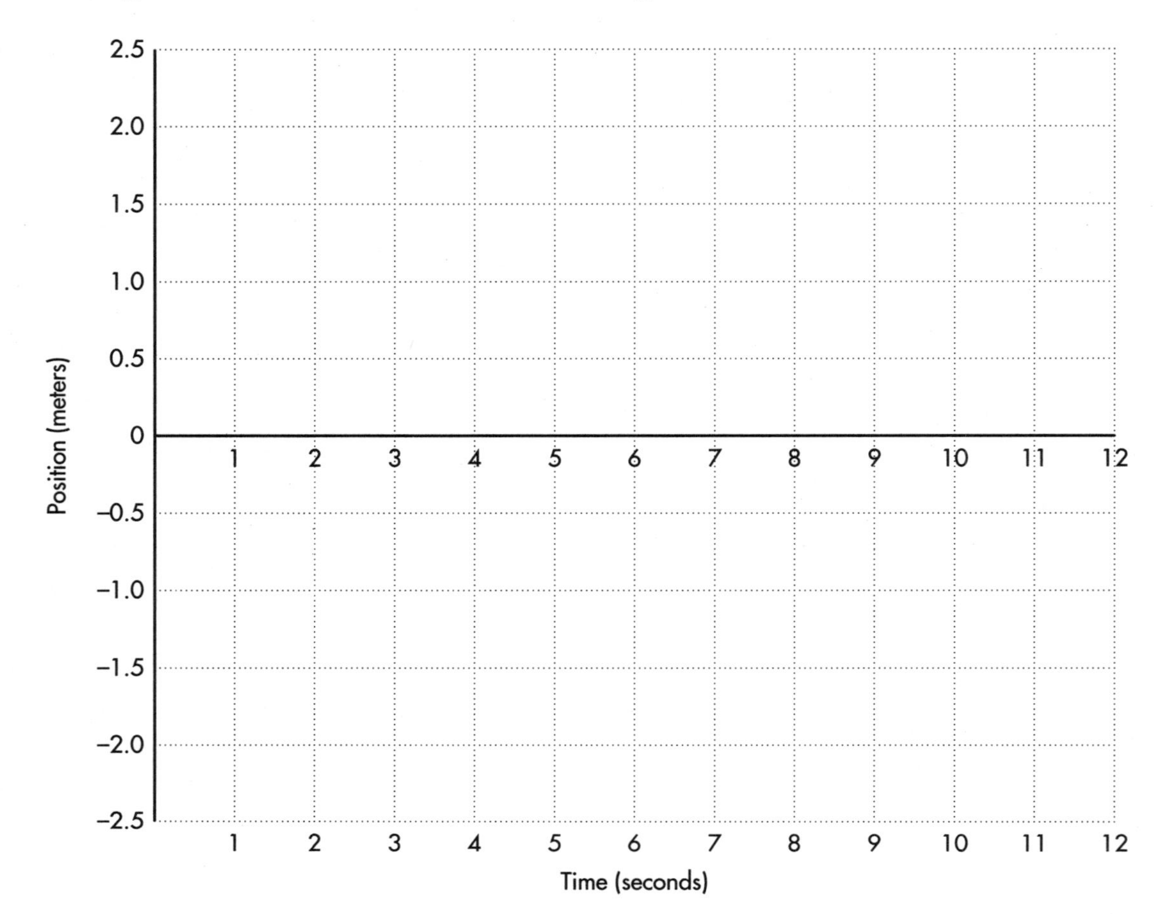

Name Date

What's the Story? (page 1 of 2)

Cut out these graphs. Match them with the stories on the next page.

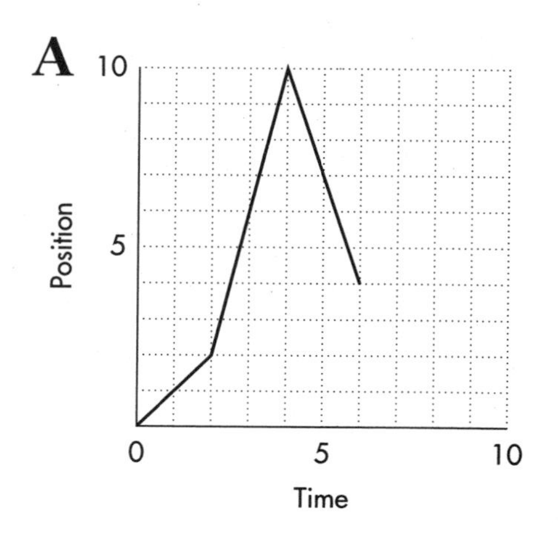

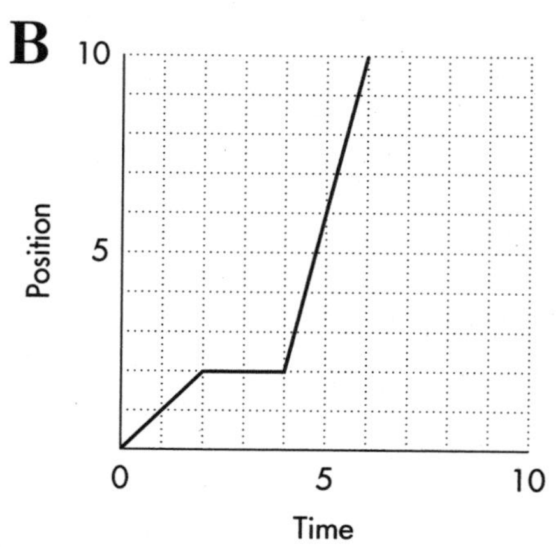

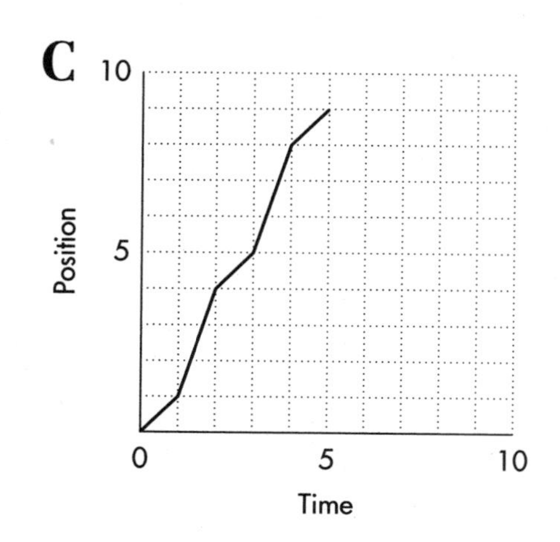

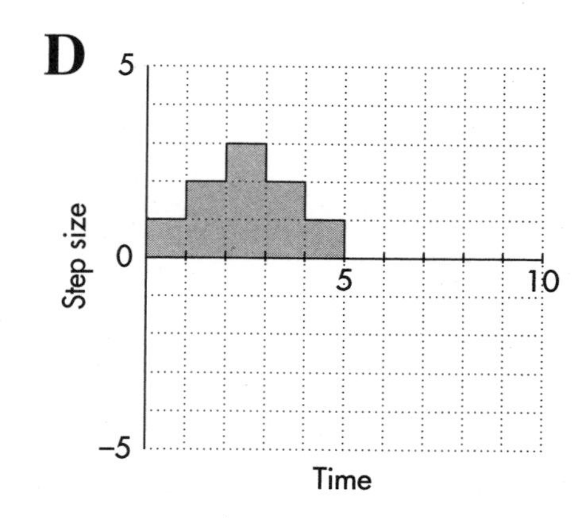

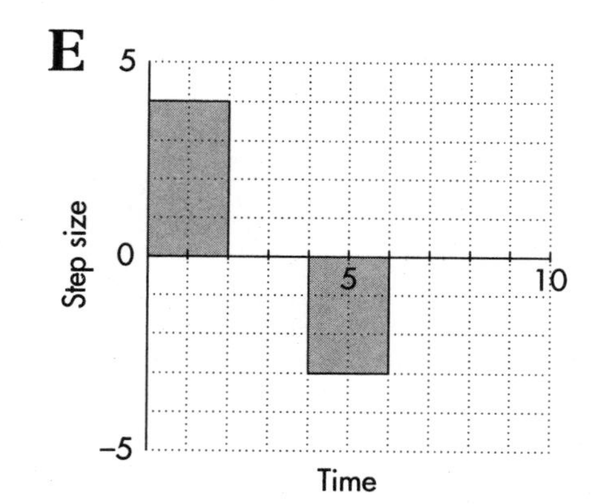

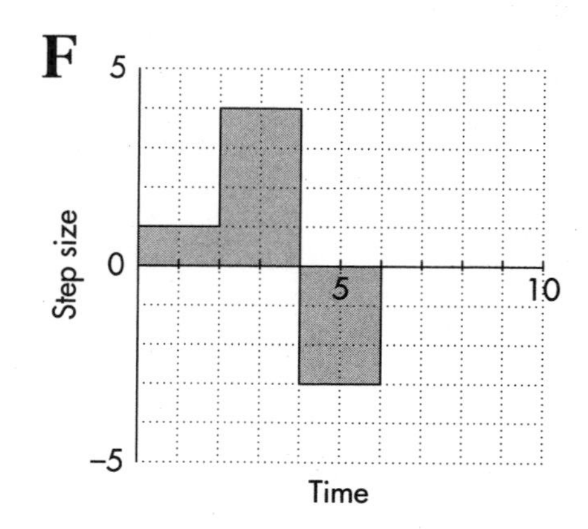

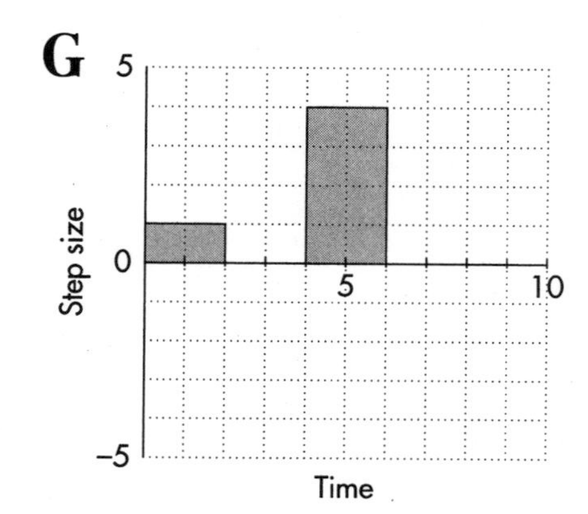

Name Date

Student Sheet 19

What's the Story? (page 2 of 2)

Below each story, paste the two graphs that fit the story. Write how you know *those* graphs go with *that* story.

Story 1 I went very slowly for 2 seconds, then stopped for 2 seconds, then went very fast to the end.

Story 2 I started slowly and went faster and faster. Then I went slower and slower.

Story 3 I went slowly for a while and then fast to the end. Then I turned around and came part of the way back in the other direction.

ONE-CENTIMETER GRAPH PAPER

Graph Shapes

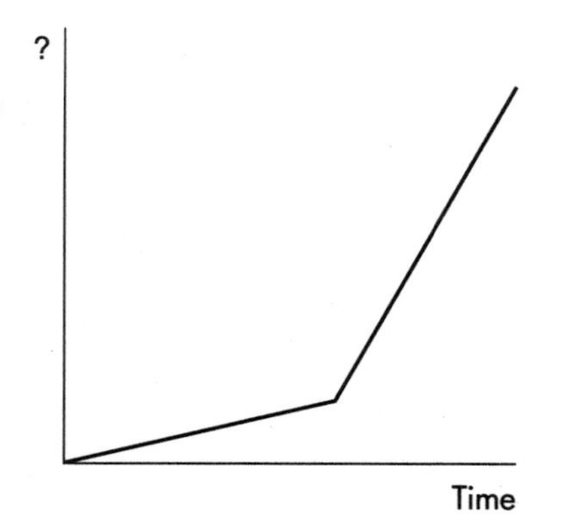

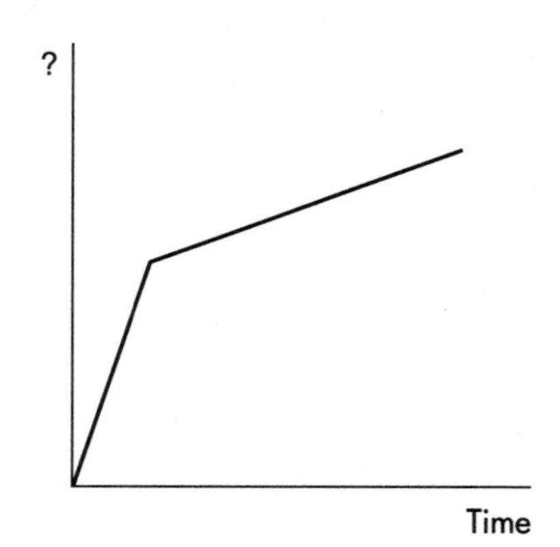

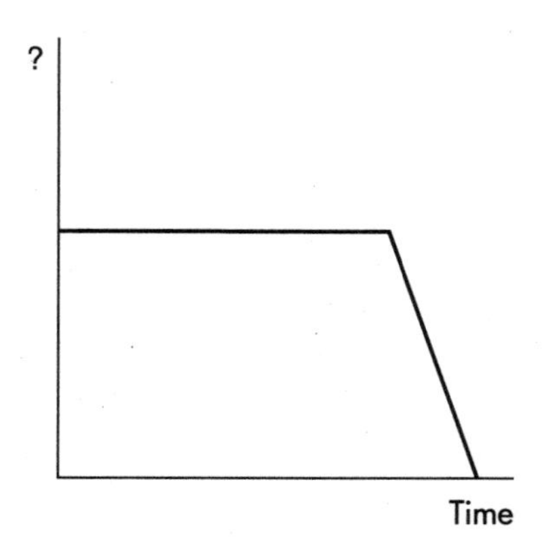

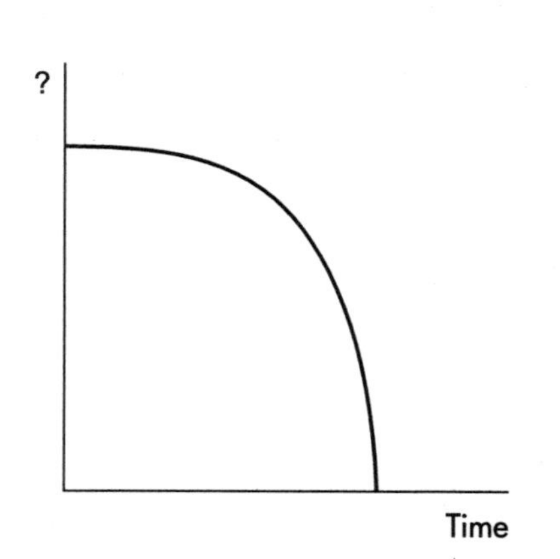

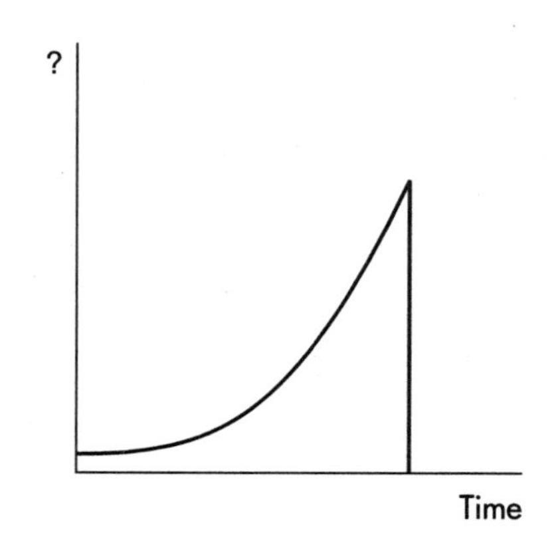

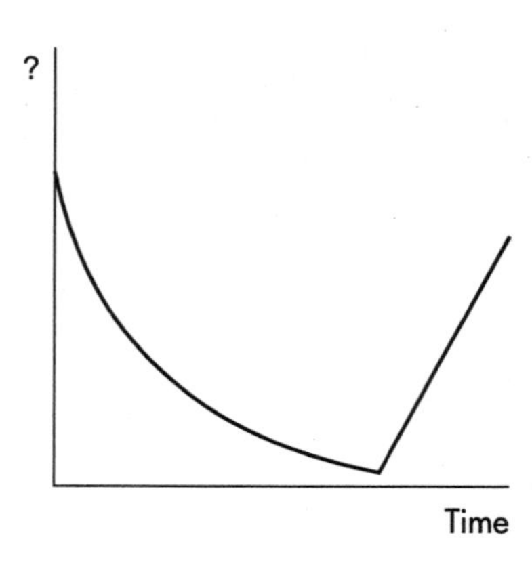

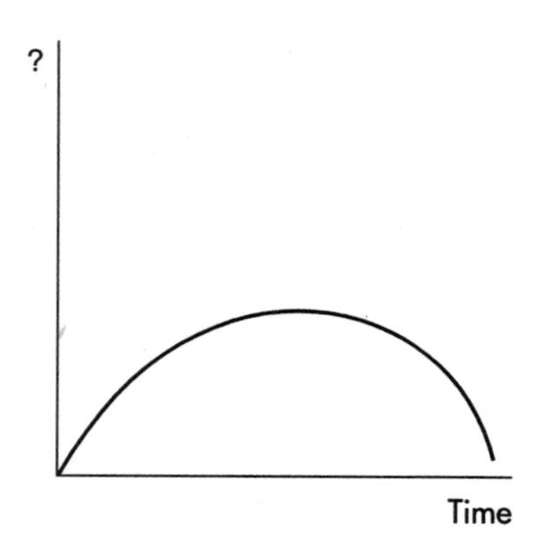

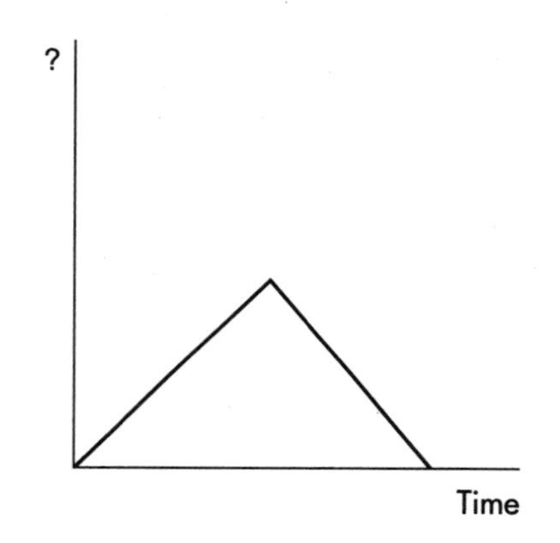

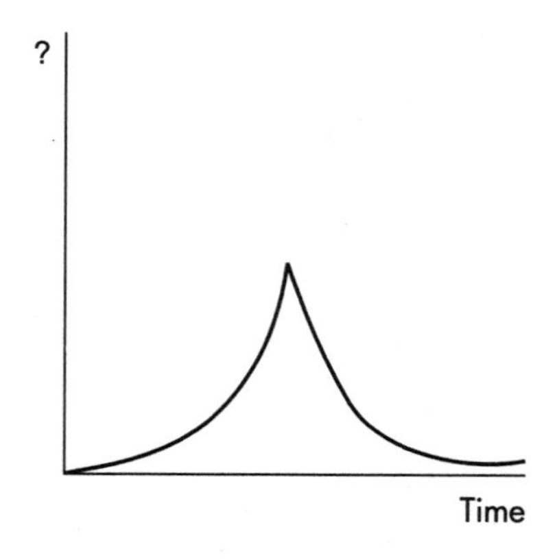

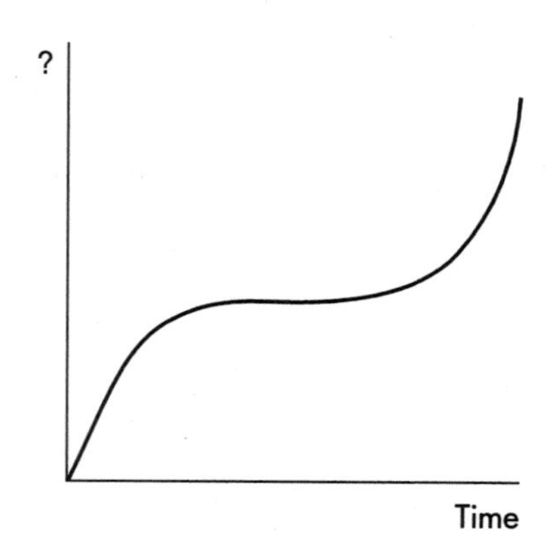

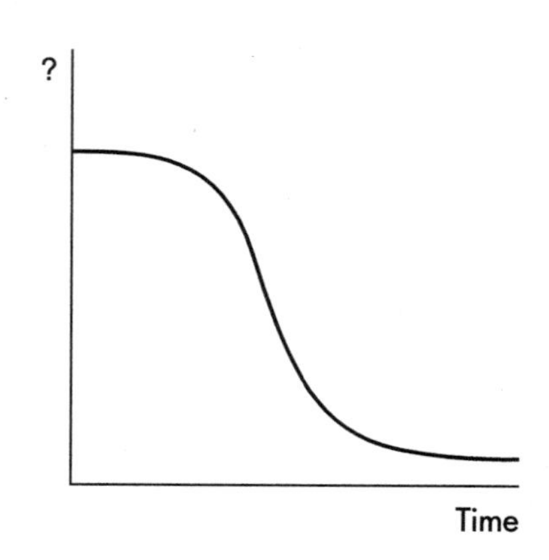

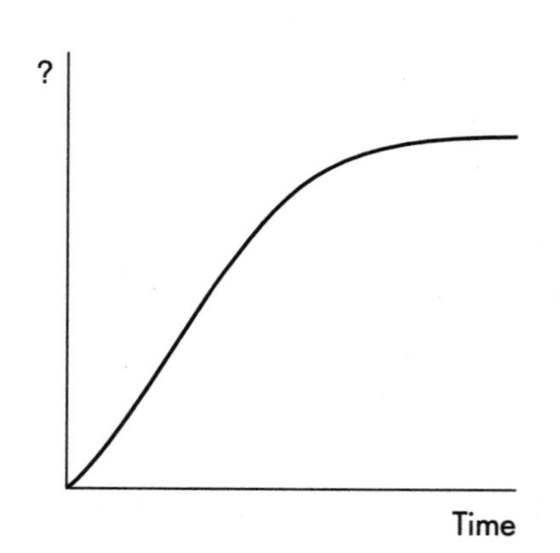